An Overview of Acquisition Reform Cost Savings Estimates

MARK LORELL
JOHN C. GRASER

Prepared for the United States Air Force

Project AIR FORCE
RAND

Approved for public release; distribution unlimited

The research reported here was sponsored by the United States Air Force under Contract F49642-01-C-0003. Further information may be obtained from the Strategic Planning Division, Directorate of Plans, Hq USAF.

Library of Congress Cataloging-in-Publication Data

Lorell, Mark A., 1947-
  An overview of acquisition reform cost savings estimates / Mark A. Lorell, John C. Graser.
      p. cm.
  Includes bibliographical references.
  "MR-1329."
  ISBN 0-8330-3018-3
  1. United States. Air Force—Procurement—Estimates. 2. Airplanes, Military—United States—Costs. I. Graser, John C. II. Title.

UG1123 .L67 2001
358.4'16212—dc21

2001031895

RAND is a nonprofit institution that helps improve policy and decisionmaking through research and analysis. RAND® is a registered trademark. RAND's publications do not necessarily reflect the opinions or policies of its research sponsors.

Cover design by Eileen Delson La Russo

© Copyright 2001 RAND

All rights reserved. No part of this book may be reproduced in any form by any electronic or mechanical means (including photocopying, recording, or information storage and retrieval) without permission in writing from RAND.

Published 2001 by RAND
1700 Main Street, P.O. Box 2138, Santa Monica, CA 90407-2138
1200 South Hayes Street, Arlington, VA 22202-5050
201 North Craig Street, Suite 102, Pittsburgh, PA 15213
RAND URL: http://www.rand.org/
To order RAND documents or to obtain additional information,
contact Distribution Services: Telephone: (310) 451-7002;
Fax: (310) 451-6915; Email: order@rand.org

PREFACE

In support of the project "The Cost of Future Military Aircraft: Historical Cost Estimating Relationships and Cost Reduction Initiatives," this study gives an overview of a wide range of published estimates and projections of potential cost savings that are attributed to a variety of weapon system acquisition reform (AR) measures. These estimates are compared in accordance with a taxonomy developed by the authors. Although the origins and quality of the estimates are discussed, no independent estimates have been generated, nor are the existing published estimates analyzed in depth. Rather, the authors present a taxonomy of current AR initiatives; review published estimates of the cost savings attributed to these initiatives; report the views of industry and government officials on the potential cost savings from AR; and discuss the structuring and implementation of programmatic AR measures based on lessons learned from existing AR pilot programs.

The project is in the RAND Project AIR FORCE Resource Management Program. The research is sponsored by the Principal Deputy, Office of the Assistant Secretary of the Air Force (Acquisition), and by the Office of the Technical Director, Air Force Cost Analysis Agency.

This study should be of interest to government and industry officials concerned with assessing the potential cost savings that current AR measures may generate when applied to major weapon system R&D and procurement programs.

The information collection cutoff date was December 1999.

PROJECT AIR FORCE

Project AIR FORCE, a division of RAND, is the Air Force federally funded research and development center (FFRDC) for studies and analysis. It provides the Air Force with independent analyses of policy alternatives affecting the development, employment, combat readiness, and support of current and future aerospace forces. Research is performed in four programs: Aerospace Force Development; Manpower, Personnel, and Training; Resource Management; and Strategy and Doctrine.

CONTENTS

PREFACE	iii
FIGURE	vii
TABLES	ix
SUMMARY	xi
ACKNOWLEDGMENTS	xxvii
ACRONYMS	xxxi

Chapter One
    INTRODUCTION AND ACQUISITION REFORM
        MEASURES TAXONOMY 1

Chapter Two
    DoD REGULATORY AND OVERSIGHT COMPLIANCE
        COST PREMIUM 9
    Introduction 9
    Defining the Regulatory and Oversight Compliance Cost
        Premium 10
    The Coopers & Lybrand Study 15
    Early Results of DoD Initiatives to Reduce the DoD
        Regulatory and Oversight Burden 18
    Early Non-DoD Assessments of Overall DoD
        AR Savings 22
    Additional Observations from Industry and Government
        Interviews 24
    Oversight Compliance Cost Premium Summary
        and Cost Savings Observations 29

Chapter Three
COMMERCIAL-LIKE PROGRAM STRUCTURE 31
Introduction 31

Chapter Four
THREE U.S. AIR FORCE ACQUISITION REFORM PILOT MUNITIONS PROGRAMS 39
Munitions Program Overviews 39
Summary of Air Force AR Munitions Programs Cost Savings 80

Chapter Five
OTHER COMMERCIAL-LIKE AR PILOT PROGRAMS 83
Introduction 83
Selected U.S. Air Force Space AR Lead Programs: SBIRS and EELV 84
Selected Defense Acquisition Pilot Programs (DAPPs) . . . 92
Selected DARPA Section 845 "Other Transaction" Programs 96
Additional Observations from Industry and Government Interviews 100
Summary Observations on Commercial-Like AR Pilot Programs 106

Chapter Six
MULTIYEAR PROCUREMENT 109

Chapter Seven
CONCLUSIONS ON AR COST SAVINGS ESTIMATES 115
Summary of the DoD Regulatory and Oversight Cost Premium Estimates 116
Summary of Savings from AR Pilot and Demonstration Programs (Commercial-Like Program Structure) . . . 119
Summary of Multiyear Procurement Savings Estimates . . 121
Concluding Observations: Some Rules of Thumb for Cost Estimators 121

Appendix
A. Subjects of the Three RAND Studies on Industry Initiatives Designed to Reduce the Cost of Producing Military Aircraft 125
B. Acquisition Reform Cost Questions 129

BIBLIOGRAPHY 135

FIGURE

4.1. JDAM Baseline Weapons 41

TABLES

S.1.	A Taxonomy of AR Measures	xv
S.2.	Early Subjective Estimates of the DoD Regulatory and Oversight Compliance Cost Premium	xvii
S.3.	Data-Based Estimates of the DoD Regulatory and Oversight Compliance Cost Premium	xviii
S.4.	Summary of Initial Assessments of Overall DoD AR Savings (in percentages)	xix
S.5.	Summary of Savings from AR Pilot and Demonstration Programs (in percentages)	xxii
S.6.	Summary of Multiyear Procurement Savings Estimates	xxiv
1.1.	A Taxonomy of AR Measures	5
2.1.	Early Subjective Estimates of the DoD Regulatory and Oversight Cost Premium	12
2.2.	DoD Regulatory Compliance Cost Premium: C&L Top Ten Cost Drivers	16
3.1.	Elements of a Commercial-Like Program Structure	33
4.1.	Commercial/Military Mix of JDAM Contractor Production Lines	49
4.2.	First Official Pre-DAPP JDAM Program Cost and AUPP Estimates, August 1993	66
4.3.	Pre- and Post-DAPP JDAM Program Costs and AUPP	67
4.4.	Pre- and Post-DAPP JDAM Development Cost and AUPP	68
4.5.	Estimated EMD and Production Percentage AR Savings for JDAM	69

4.6.	Estimated EMD and Production Percentage AR Savings for WCMD by Category	70
4.7.	Estimated JASSM R&D and Production Percentage AR Savings	72
4.8.	Summary of Estimated Percentage Savings for U.S. Air Force Munitions AR Lead Programs	72
5.1.	Selected Air Force Space AR Pilot Program Estimated Savings	87
5.2.	FSCATT: Breakdown of Estimated AR Program Savings	93
5.3.	Projected FSCATT AR Savings	94
5.4.	JPATS Program Cost Estimates	95
5.5.	FY97 POE Projected JPATS and Other Savings as Percentage of FY92 and FY95 POEs	96
5.6.	DARPA Section 845 OTA HAE UAV R&D Programs	100
6.1.	Selected Multiyear Procurement Savings Estimates	111
7.1.	Early ROM Estimates of the DoD Regulatory and Oversight Cost Premium	117
7.2.	Estimates of the DoD Regulatory and Oversight Compliance Cost Premium	118
7.3.	Summary of Initial Assessments of Overall DoD AR Savings	118
7.4.	Summary of Savings from AR Pilot and Demonstration Programs (Commercial-Like Program Structure)	120
7.5.	Summary of Multiyear Procurement Savings Estimates	121

SUMMARY

INTRODUCTION AND ACQUISITION REFORM MEASURES TAXONOMY

This report supports a larger RAND project entitled "The Cost of Future Military Aircraft: Historical Cost Estimating Relationships and Cost Reduction Initiatives." The purpose of the project is to update the technical cost models and cost estimating relationships (CERs) for fixed-wing combat aircraft in light of R&D, manufacturing, organizational, and programmatic advances and reforms that have taken place over the past decade. Taking into account the potential overlap of claimed savings resulting from new (post-1990) aircraft design and manufacturing initiatives (especially for advanced airframe materials), acquisition reform, and lean implementation, the RAND project divided the research effort into five areas:[1]

1. New fabrication and assembly processes related to advanced airframe materials;[2]

2. Government changes in acquisition processes or changes in the relationship between the government and Department of Defense (DoD) prime contractors, generally included under the rubric of "acquisition reform";

---

[1] See Appendix A for a listing of all military aircraft initiatives addressed in three of these reports.

[2] Obaid Younossi, Michael Kennedy, and John C. Graser, *Military Airframe Costs: The Effects of Advanced Materials and Manufacturing Processes*, Santa Monica: RAND, MR-1370-AF, 2001.

3. Lean implementation and other initiatives oriented primarily toward processes within a prime airframe manufacturer or relationships between these primes and their suppliers;[3]
4. Technology and process improvements in military avionics development and manufacturing, especially as they relate to the Joint Strike Fighter (JSF);[4] and
5. Technology and process improvements in military aircraft engines (research in progress).

This report covers research on acquisition reform (AR). Its purpose is to determine whether published estimates in the literature are sufficiently robust to contribute to the development of adjustment factors for use in predictive cost models that reflect the effects of AR on the costs of developing and producing fixed-wing combat aircraft. The report reviews a wide range of published estimates and projections of claimed savings that may arise from a variety of weapon system AR measures. However, no independent RAND estimates of potential AR cost savings have been generated, nor have any of the published estimates been analyzed. Rather, the existing estimates are grouped into logical categories and compared, and the variations, historical origins, and relative quality of these estimates are discussed. In addition, the report presents the views of numerous prime contractors on potential AR cost savings, all derived from a series of interviews conducted in 1998.[5]

---

[3]Cynthia R. Cook and John C. Graser, *Military Airframe Acquisition Costs: The Effects of Lean Manufacturing*, Santa Monica: RAND, MR-1325-AF, 2001.

[4]Mel Eisman, Jon Grossman, Joel Kvitky, Mark Lorell, Phillip Feldman, Gail Halverson, and Andrea Mejia, *The Cost of Future Military Aircraft Avionics Systems: Cost Estimating Relationships and Cost Reduction Initiatives*, Santa Monica: RAND, limited document, not for public distribution, 2001.

[5]Industry sites visited by RAND include Boeing Military Aircraft and Commercial Aircraft, Seattle, Washington; Boeing McDonnell Military Aircraft and Missile Systems, St. Louis, Missouri; British Aerospace Military Aircraft and Aerostructures, Warton and Samlesbury, United Kingdom; DaimlerChrysler Aerospace Airbus GmbH, Bremen, Germany; DaimlerChrysler Aerospace AG Military Aircraft, Munich, Germany; Lockheed Martin Aeronautical Systems, Marietta, Georgia; Lockheed Martin Skunk Works, Palmdale, California; Lockheed Martin Tactical Aircraft Systems, Fort Worth, Texas; Northrop Grumman Electronic Sensors and Systems Sector, Baltimore, Maryland; Northrop Grumman Integrated Systems and Aerostructures, Air Combat Systems, El Segundo, California; Northrop Grumman Integrated Systems and Aerostructures, Dallas, Texas; Northrop Grumman Integrated Systems and

The report concludes that on the whole, there is insufficient evidence in the published literature to support the development of precise adjustment factors for AR cost savings that can be used with confidence in technical cost models for military combat aircraft. At the same time, our research suggests that at least in some categories of AR measures, rough order-of-magnitude (ROM) estimates or "rules of thumb" for potential AR cost savings can be developed that may be of use to cost estimators in limited circumstances.

There is a vast body of literature on AR that covers a wide variety of measures. As a result, our first task was to develop a taxonomy of AR measures that would provide a rational ordering and coherent linkage between these various measures. Table S.1 presents our taxonomy of current major AR measures and initiatives, which was developed for the purpose of assigning published cost savings estimates to specific elements. As indicated in Table S.1, we suggest three major AR categories: (1) reducing regulatory and oversight burden; (2) commercial-like program structure; and (3) multiyear procurement. Table S.1 also presents subelements of the second category together with suggestions on how these subelements might be linked to the main category.

The tables that follow summarize the data presented in this report on published AR cost savings estimates and projections. The many assumptions underlying each estimate and the numerous caveats included in the body of this report are not repeated here. It is important to note, however, that these estimates vary considerably in both quality and methodology and must therefore be used with caution. A detailed reading of the main text of this report is necessary to clarify the many limitations and caveats that must be applied in their use.

Most of the following tables have a column labeled "estimate quality." This column distinguishes between three types of estimates. The highest-quality estimate, labeled "actuals," signifies that the estimate of AR savings was based on actual R&D and production cost data from the specific item under consideration, compared to earlier actuals for the program prior to the imposition of acquisition re-

---

Aerostructures, Hawthorne, California; Raytheon Aircraft, Wichita, Kansas; Raytheon Sensors and Electronic Systems, El Segundo, California; and Scaled Composites, Inc., Mojave, California.

forms. Virtually none of the estimates available during the course of this research effort was of this type.[6] The second-highest-quality estimate, labeled "forecast," refers primarily to a narrow set of cases in which actual production costs for the specific article are well known but the program is being restructured in a way that is expected to reduce costs. This applies mainly to estimates of multiyear production contract savings. The third-highest-quality estimates, labeled "analysis," are made in situations where no actual costs are available for the specific item. In such cases, the anticipated pre-AR cost of a specific item, which has not yet been fully developed or entered into production, is compared to the expected cost of that item after the imposition of AR—in other words, neither the actual cost of the item under the old system nor the actual cost of the item after the imposition of AR is known. This type of estimate is based on rational analysis, past experience, data from analogous military or commercial programs adjusted to the system under examination, expert opinion, or similar methods.

Almost all the AR cost savings estimates collected in this report fall into the category of "analysis." That is, they are not based on actual data for the specific system or program structure in question, either before or after AR. This is another key reason these estimates must be treated with extreme care.

SUMMARY OF THE DoD REGULATORY AND OVERSIGHT COMPLIANCE COST PREMIUM ESTIMATES

The DoD regulatory and oversight compliance cost premium refers to the additional costs that the DoD is alleged to pay to contractors to cover the added cost of complying with the vast array of regulations and requirements imposed on the contractor by the government. This cost is alleged to be over and above what the same item would

---

[6]See the subsequent discussion on actuals.

Table S.1

A Taxonomy of AR Measures

1. **Reducing Regulatory and Oversight Compliance Cost Premium**
2. **Commercial-Like Program Structure**
   A. Emphasis on CAIV[a] through the use of:
      (1) Unit price thresholds, unit price targets
      (2) Production price requirement and commitment curves + carrots/sticks in final down-select and in production contract (including warranties, etc.)
      (3) Competition
   B. Enable CAIV through emphasis on:
      (1) Requirements reform
         (a) No "overdesigning"[b]
         (b) Prioritized tradable performance/mission requirements (threshold requirements, etc.)
      (2) Contractor configuration control, design flexibility
      (3) Commercial insertion/dual use,[c] which is made possible by
         (a) Mil spec reform[d]
         (b) Government-industry IPTs[e]
3. **Multiyear Procurement[f]**

[a]CAIV is an acronym for "cost as an independent variable." The basic concept of CAIV is that it raises rigorous production-unit cost goals to the same priority level as performance and other key system goals during the design and development phases of a weapon system. As such, it is similar to the "must cost" goals that commercial aircraft transport developers and other commercial firms impose on their designers, engineers, and subcontractors when they initiate the development of a new system. More is said on the CAIV concept in subsequent sections of this report.

[b]A more familiar term that could have been used in this context is "gold plating." This term was rejected, however, because some observers associate it with less-than-objective journalistic critiques of the defense acquisition process. The term "overdesigning" as used here means to design into a weapon system capabilities or attributes that may not be worth the extra expense or that are not essential to meeting the mission requirements.

[c]"Commercial insertion" refers to the use of commercial off-the-shelf (COTS) technologies, processes, parts, components, subsystems, and/or systems in weapon systems. The term also refers to the use of "ruggedized" or "militarized" COTS products. "Ruggedization" signifies the special packaging or other hardening of COTS products to permit them to function in harsh military environments. "Dual use" refers to technologies, manufacturing facilities, and products that are known to have or may have both commercial and military applications.

[d]"Mil spec" is an acronym for military specifications and standards.

[e]IPT = Integrated Product Team.

[f]"Multiyear procurement" refers to government authorization for the procurement of specific numbers of production systems beyond the normal single-year government procurement funding cycle. Multiyear procurement requires special congressional approval.

cost were it acquired in a purely commercial environment by a civilian customer. Table S.2 summarizes several late-1980s and early-1990s DoD cost premium estimates that typify those used by early advocates of AR.

As Table S.2 indicates, early estimates vary considerably in quality and methodology, and none are directly comparable. Most are based on expert opinion, anecdotal information, or projections derived from commercial analogies that may or may not be appropriate. For the most part, such estimates could thus be characterized as informed guesses. Some of these estimates include potential cost savings from factors other than the reduction in compliance costs, such as cost benefits gained from using commercial technologies and parts. However, it is not always clear whether such factors are included in the estimates.

Table S.3 summarizes the most important estimates of the DoD regulatory and oversight compliance cost premium. These estimates are based on actual data derived from Coopers & Lybrand (C&L) and other studies conducted during the initial phases of the current AR reform effort. It should be noted, however, that these estimates are based on limited data and on varying methodologies. In addition, the methodologies they employ are not always fully transparent and may be open to criticism. Moreover, the raw data on which the estimates are based are seldom available. To be fully understood, this table thus requires a full reading of the main text of this report.

We believe that the most reliable of the studies outlined in Table S.3 suggest potential savings from DoD regulatory and oversight relief in the range of 1 to 6 percent. We further suggest that this range, with an average of 3.5 percent, is a reasonable ROM or "rule-of-thumb" estimate for potential savings from eliminating the DoD regulatory and oversight compliance cost premium. If one is to obtain the full benefit of savings from regulatory and oversight relief, however, the consensus view is that virtually all burdensome regulations and

Table S.2

Early Subjective Estimates of the DoD Regulatory and Oversight Compliance Cost Premium[a]

Study	Date	Estimated DoD Cost Premium/Potential Cost Savings (%)
Honeywell defense acquisition study (20 programs, contractor costs)	1986	13
RAND OSD regulatory cost study (total program costs)	1988	5–10
OTA industrial base study (total DoD acquisition budget)	1989	10–50
CSIS CMI study[b] (cost premium on identical items)	1991	30
Carnegie Commission (total DoD acquisition budget)	1992	40
ADPA cost premium study (product cost)	1992	30–50

[a]The full titles of these studies are as follows: *Defense Acquisition Improvement Study*, Honeywell, May 1986; G. K. Smith et al., *A Preliminary Perspective on Regulatory Activities and Effects in Weapons Acquisition*, Santa Monica: RAND, R-3578-ACQ, March 1988; Office of Technology Assessment, *Holding the Edge: Maintaining the Defense Technology Base*, Vol. II Appendix, Washington, D.C.: USGPO, April 1989; *Integrating Commercial and Military Technologies for National Security: An Agenda for Change*, Washington, D.C.: Center for Strategic and International Studies, April 1991; *A Radical Reform of the Defense Acquisition System*, Carnegie Commission on Science, Technology, and Government, December 1, 1992; and *Doing Business with DoD—The Cost Premium*, Washington, D.C.: American Defense Preparedness Association, 1992.
[b]CMI = Civil-military integration.

oversight must be removed from all programs and by all government customers for each major government contractor or contractor facility. Because of these limitations and caveats, it is probably not appropriate to use 3.5 percent as a technical adjustment factor in mathematical models that employ empirically tested CERs.[7]

---

[7]See Concluding Observations in Chapter Seven of this report.

Table S.3
Data-Based Estimates of the DoD Regulatory and Oversight Compliance Cost Premium

Study or Program and Date[a]	C&L Top 10 Cost Drivers (%)	C&L Top 24 Cost Drivers (%)	Overall Cost Premium or Savings Potential (%)	Estimate Quality
C&L (1994)	8.5	13.4	18	Forecast
NORCOM (1994)			27	Forecast
DoD Regulatory Cost Premium Working Group (1996)		6.3		Forecast
DoD Reinvention Lab (1996)	1.2–6.1			Forecast
SPI (1998)			0.5	Limited actuals
WCMD (1996) (CDRLs only)			3.5 (R&D)	Analysis
FSCATT (1995)			2	Analysis
B-2 Upgrade (CDRLs only)			2.3	Forecast

[a]SPI = Single-Process Initiative; WCMD = Wind-Corrected Munitions Dispenser; CDRL = Contractor Data Requirements List; FSCATT = Fire Support Combined Arms Tactical Trainer.

This report also examines nongovernment and General Accounting Office (GAO) estimates of overall DoD AR program savings from the early stages of the Clinton administration reform efforts (see Table S.4). These studies are based largely on comparisons of overall program budget data and on projections from different fiscal years or periods. For the most part, they offer little or no breakout of specific AR measures or of how and to what extent such measures might have contributed to the changes in estimates. It is not unreasonable to assume that most of the reported actual savings (as opposed to the reported future cost avoidance beyond FY01) was due to reductions in the DoD regulatory and oversight burden.[8] We conclude this for two reasons:

- Most of the programs examined for these estimates and projections had been under way for some time as traditional programs before AR; and
- More radical programmatic acquisition reforms had not been fully implemented at the time the studies collected data.

Although these estimates are not directly comparable either to each other or to earlier estimates of the potential DoD regulatory and oversight reform cost savings, we believe that they add some support to the notion that the DoD regulatory and oversight cost burden is in the range of 1 to 6 percent.

Table S.4

Summary of Initial Assessments of Overall DoD AR Savings (in percentages)

Study and Date	FY95–FY01	1996	FY95–FY02	Estimate Quality
RAND (1996)	4.4			Forecast
MIT (1997) (average of 23 MDAPs)[a]		4.3		Forecast
GAO (1997) (average of 33 MDAPs)			-2^b	Forecast
GAO (1997) (average of 10 MDAPs with cost savings)			4	Forecast

[a]MDAP = Major Defense Acquisition Program.
[b]This estimate does not dispute the existence of cost savings from AR for these programs. Rather, it suggests that on average, cost savings are often offset by cost increases elsewhere or by reinvestment.

---

[8]The projections of future cost avoidance are obviously just estimates based on past experience.

Given the uncertainties and ambiguities inherent in the data available on DoD regulatory and oversight cost savings, how should cost estimators use this information? We have concluded that it is reasonable to assume program savings of 3 to 4 percent due to reductions in the regulatory and oversight burden. In other words, if one is using a pre-AR (pre-1994) program as an estimating analogy for a similar new program, it is reasonable to assume cost reductions at the program acquisition level of 3 to 4 percent due to reductions in the regulatory and oversight burden. However, if the cost analysis is developed using prior program direct or indirect labor hours, most of the AR savings from reductions in regulatory and oversight burdens should already be reflected in the negotiated forward pricing rate agreements (wrap rates), so no further adjustment would be warranted in the rates themselves. This is because most regulatory burden cost savings are in the area of indirect costs and should thus show up in overhead cost savings. Because AR has been in existence since 1995, most of the realizable reductions in regulatory and oversight burdens should already have been calculated between the contractor and the Defense Contract Management Agency (DCMA). This assumes, however, that a comprehensive program of relief from DoD regulatory and reporting requirements has been applied to all the programs of a specific contractor or to all the programs at a specific facility. This, of course, is not actually the case.

AR reductions between suppliers and the prime may have to be assessed separately, as factors such as regulatory flow-down and the cost effects of strategic supplier relationships must be taken into account. Although AR has focused mainly on interactions between the government and the primes, there may be areas between primes, subcontractors, and suppliers that result in further savings due to reductions in regulatory and oversight burdens.

SUMMARY OF SAVINGS FROM COMMERCIAL-LIKE AR PILOT PROGRAMS

Commercial-like AR pilot programs exhibit a complex mixture of the numerous reform measures that are outlined in Table S.1 and discussed in detail in the body of this report. The purpose of these measures is to structure weapon system acquisition programs so that the incentives provided to contractors are more like those found in commercial R&D and production programs. These measures seek to incentivize the contractor to focus on cost as a primary objective and to use commercial standards, technology, parts, and components.

It is critical to note that the claimed savings from these programs are based on comparing estimated projected costs before the imposition of AR measures with estimated projections following the imposition of AR measures. Few are based on hard data. That is, few of the estimates contain actuals, or actual cost data based on real work undertaken during product development and production. Most of the estimates were made before the beginning of system development or in the early phases of development. Even in cases where actuals were used in order to show claimed AR savings, the actuals were compared to an earlier estimate that was only a forecast and that itself was not based on actuals (i.e., on the actual pre-AR costs of the item). These estimates must therefore be viewed with extreme caution. Table S.5 summarizes the cost savings estimates from these programs.

The data in Table S.5 suggest that R&D savings in the range of 15 to 35 percent may be possible in programs that are fully restructured in a commercial-like manner in accordance with the concepts of cost as an independent variable (CAIV), as discussed in great detail in the body of this report. The likely scale of anticipated production savings is much more uncertain. However, the three best-documented cases—Joint Direct Attack Munition (JDAM), Wind-Corrected Munitions Dispenser (WCMD), and Joint Air-to-Surface Standoff Missile (JASSM)—suggest that savings of up to 65 percent are possible, at least in programs for less complex systems with high production runs.

Some additional qualifications must be noted in discussing these outcomes. First, the reforms used in these pilot programs have not been widely used as an integrated package outside these AR demonstration programs. Furthermore, many AR pilot programs are relatively small and are characterized by low technological risk, commercial derivative items, and large production runs. Thus, the scale of potential cost benefits for a large, complex weapon system that employs high-risk, cutting-edge technology remains uncertain. Finally and most significantly, several of these programs have only

Table S.5
Summary of Savings from AR Pilot and Demonstration Programs[a]
(in percentages)

Program[b]	Program Savings (%)	R&D Savings (%)	Production Savings (%)	Estimate Quality
JDAM		15	60	Forecast
WCMD		35	64	Forecast
JASSM	44[b]	29	31	Analysis
EELV		20–33	25–50	Analysis
SBIRS		15		Analysis
FSCATT	13.5	16–34	7	Analysis
JPATS	18.9[c]	13.6	–26.6[d]	Analysis
Tier III-		20		Analysis
Tier II+		3		Analysis
ASP		30		Analysis
AAAV			10–20	Analysis

[a]Note the important qualifications explained in main text.
[b]EELV = Evolved Expendable Launch Vehicle; SBIRS = Space-Based Infrared System; JPATS = Joint Primary Aircraft Training System; ASP = Arsenal Ship Program, AAAV = Advanced Amphibious Assault Vehicle.
[c]Overall program cost savings claimed by the DoD, March 1999.
[d]Despite a large increase in production costs, overall program costs declined significantly because of a large anticipated reduction in operations and support (O&S) costs. In March 1999, the DoD claimed an overall JPATS contract cost savings of 49 percent.

recently entered the low-rate initial production (LRIP) stage; the majority have not even completed engineering and manufacturing development (EMD).

Our detailed review of several AR pilot programs, as well as the consensus views we gleaned from extensive RAND interviews with industry and government representatives, provided additional insights regarding cost savings from the commercial-style program structures discussed above:

- Requirements reform (performance-based specifications) and CAIV ("must-cost" objectives used during EMD in the downselect decision) are crucial for cost savings. CAIV essentially entails a trade-off of technical capabilities against cost. The key to CAIV is avoiding "overdesigning" and retaining only mission-essential capabilities.

- Maximizing the use of commercial parts and technology in weapon systems to the extent that it does not compromise critical system performance capabilities has a high AR savings potential, especially in electronics.
- Requirements reform, regulatory reform, CAIV, and especially contractor configuration control are all necessary to motivate greater use of commercial parts and technology by contractors.
- Commercial-style programs with greater contractor cost sharing would be encouraged by reducing constraints on foreign sales and technology transfer.
- Commercial-like "must-cost" pricing goals combined with competition appear to incentivize contractors to control costs.
- Commercial-style R&D and production programs with contractor configuration control may require contractor logistics support once systems are fielded. The Air Force may face serious problems applying these types of AR reforms to large, complex platform development programs.
- True dual-use (commercial and military) utilization of production facilities on a system or major-subsystem level is still rare. Government regulations and technology differences remain significant barriers.
- The level of AR actually implemented on some government pilot programs has been less than some contractors had expected.

Given the lack of data and the many uncertainties and complexities that surround commercial-like AR programs, how should cost estimators deal with such programs? It is our view that if an acquisition program entails extensive civil-military integration (CMI) and insertion of COTS parts and technology, specific cost reductions need to be assessed as appropriate, probably at the purchased-materials and purchased-parts levels of a cost estimate. For programs such as JDAM and various avionics efforts that claim large savings from AR, vendor-supplied parts, components, boards, and the like account for as much as 80 to 90 percent of recurring costs. Yet there can be wide variations from one system or program to another. Thus, no easy rule of thumb can be applied in this area.

If separate and significant AR initiatives can be identified in specific programs, they should be evaluated individually and the results used to adjust the baseline cost estimate, assuming that the baseline is derived from historical, pre-AR costs. One of the most important AR initiatives is the extensive use of CAIV. However, once the final design configuration is determined and frozen following the CAIV process, the AR savings from CAIV would already be clearly reflected in the life cycle cost (LCC) baseline of the system. However, if a program entails significant contractor configuration control throughout EMD and production, a careful assessment of ongoing cost-saving opportunities and contractor incentives is warranted. Possible positive and negative operations and support (O&S) implications of contractor configuration and Total System Performance Responsibility (TSPR) need to be examined.

Table S.6

Summary of Multiyear Procurement Savings Estimates[a]

Program[b]	Production Savings (%)	Estimate Quality
F-16 (FY82–85)	10	Forecast
F-16 (FY86–89)	10	Forecast
F-16 (FY90–93)	5.5	Forecast
F-16 (FY99–02)	5.4	Forecast
CDE for C-17	8.2	Forecast
C-17 (airframe)	5.5	Forecast
Javelin ATGM	14.3	Analysis
MTVR	7.4	Analysis
CH-60 (U.S. Navy and U.S. Army)	5.5	Forecast
DDG-51 (FY98–01)	9	Forecast
F-22 (1996 CAIG/JET)	3.9–4.7	Analysis
F/A-18E/F (target)	7.4	Analysis

[a] Savings percentages include government investments for cost reduction initiatives for C-17 airframe and F/A-18E/F.

[b] CDE = Commercial Derivative Engine; ATGM = anti-tank guided missile; MTVR = Medium Tactical Vehicle Replacement; CAIG = Cost Analysis Improvement Group; JET = Joint Estimate Team.

SUMMARY OF MULTIYEAR PROCUREMENT SAVINGS ESTIMATES

Data and analytical forecasts based on past experience suggest that multiyear contracts can save roughly 5 percent, and possibly as much as 10 percent on production contracts. Table S.6 summarizes the data and forecasts that support this claim.

Again, it is important to mention a key caveat regarding the comparisons on which these and many other savings claims are made: Such claims are based on comparing preprogram estimates of the program costs on a year-to-year contract to a multiyear basis. Once a decision is made to follow one path or the other, the two can no longer be compared on an equivalent basis, as fact-of-life changes occur throughout a production program. The savings are thus based on the best estimates available at the time of the decision, not on any actual historical data for the path not chosen.

Based on the evidence collected here, and keeping in mind the caveats stated above, we conclude that multiyear contracts that are effectively implemented by the prime contractor and government customer can be expected to produce approximately 5 percent or greater savings compared to traditional programs. Multiyear contracts permit long-range planning by contractors. In addition, they permit larger buys of materials and parts, and allow for strategic relationships between primes and subcontractors. Therefore, multiyear contracting should inherently result in some cost savings. However, strategic sourcing relationships between primes, subcontractors, and suppliers fostered under lean manufacturing will have to be evaluated by cost estimators in conjunction with the multiyear savings to ensure that double counting is avoided.[9]

---

[9]See Cook and Graser, *Military Airframe Acquisition Costs: The Effects of Lean Manufacturing,* for a discussion of strategic supplier relationships.

ACKNOWLEDGMENTS

The authors gratefully acknowledge the assistance of many government and aerospace industry officials in conducting this research effort. Without their help, this report could not have been written. Unfortunately, those who so generously contributed of their time and knowledge are far too numerous to list individually. Listed below are the principal government and industry organizations that were visited in the course of this research effort. The main points of contact for each organization are also given.

- **Office of the Assistant Secretary of the Air Force for Acquisition (SAF/AQ):** Lieutenant General George K. Muellner (ret.), General Gregory S. Martin, and Lieutenant General Stephen B. Plummer.

- **Air Force Cost Analysis Agency:** Joseph Kammerer, Director; John Dorsett, former Technical Director; and Jay Jordan, Technical Director.

- **Joint Strike Fighter Program Office:** Lieutenant General Leslie Kenne, former Director; Major General Michael Hough, Director; and Lieutenant Colonel Thomas Dupré.

- **Office of the Secretary of Defense, Program Analysis and Evaluation:** Lieutenant Colonel David Nichols.

- **U.S. Air Force Aeronautical Systems Center, F-22 System Program Office:** Donna Vogel and John Eck.

- **United Kingdom Ministry of Defence, London:** Simon Webb.

- **Federal Republic of Germany Ministry of Defense, Bonn, Germany:** Rolf Schreiber, Ministerial dirigent.

- **NETMA (NATO EF2000 and Tornado Development, Production, and Logistics Management Agency), Unterhaching, Germany:** Christian Biener, Deputy General Manager.
- **The Boeing Company, Long Beach, California:** Tim Degani and Jay Asher.
- **The Boeing Company, St. Louis, Missouri:** Donna Chapman.
- **The Boeing Company, Seattle, Washington:** Dave Brower and Jim Hayes.
- **British Aerospace Military Aircraft and Aerostructures, Warton and Samlesbury, United Kingdom:** Andrew Hodgson.
- **DaimlerChrysler Aerospace, Munich, Germany:** Hanas Ross.
- **Lockheed Martin Aeronautical Systems, Marietta, Georgia:** Byron Burgess.
- **Lockheed Martin Skunk Works, Palmdale, California:** Kevin Woodward.
- **Lockheed Martin Tactical Aircraft Systems, Fort Worth, Texas:** Lew Jobe and Dr. Bryan Tom.
- **Northrop Grumman Corporation, Dallas, Texas:** Dr. John Gilliland.
- **Northrop Grumman Corporation, El Segundo, California:** Jim Byrd.
- **Northrop Grumman Corporation, Hawthorne, California:** Jim Byrd.
- **Raytheon Aircraft, Wichita, Kansas:** Howard Stewart.
- **SAIC:** Michael A. Boito.
- **Scaled Composites, Inc., Mohave, California:** Burt Rutan.

Many RAND colleagues and other analysts also provided important comments and criticism on early drafts of this document. First and foremost, the authors wish to thank the formal reviewers of this document: Kathi Webb of RAND Washington and Dennis Smallwood, U.S. Army Military Academy, West Point. Special thanks are due to Michael Vangel, Boeing Corporation, St. Louis, for his extensive assistance with the case study of the Joint Direct Attack Munition (JDAM). Thanks are also due to two Air Force officers who provided special assistance and insights: Colonel Darrell Holcomb and Lieutenant Colonel Joe Besselman.

The authors would also like to acknowledge several Project AIR FORCE (PAF) managers whose support was crucial for the completion of this research effort: Natalie Crawford, Vice President and Director, PAF; C. Richard Neu, Associate Director, PAF; C. Robert Roll, Director, PAF Resource Management Program; and Michael Kennedy, Associate Program Director, PAF Aerospace Force Development Program. Finally, the authors acknowledge the effort's project officers: John Dorsett, former Technical Director of the Air Force Cost Analysis Agency, and Jay Jordan, current Technical Director.

ACRONYMS

AAAV	Advanced Amphibious Assault Vehicle
ABC	Activity-based costing
ACAT	Acquisition Category
ACTD	Advanced Concept Technology Demonstration
ADPA	American Defense Preparedness Association
AFMC	Air Force Material Command
AGM	Air-to-ground missile
AIM	Air Intercept Missile
AMC	Army Material Command
AOG	Aircraft on ground
AR	Acquisition reform
ASP	Arsenal Ship Program
ATGM	Anti-tank guided missile
ATR	Autonomous target recognition
AUPP	Average unit procurement price
AUPPC	Average unit production price commitment
AUPPR	Average unit procurement price requirement

BLU	Bomb, Live Unit
BMD	Ballistic missile defense
BY	Base year
C&L	Coopers & Lybrand
C/SCS	Cost/Schedule Control System
C/SCSC	Cost/Schedule Control System Criteria
CAID	Clear Accountability in Design
CAIG	Cost Analysis Improvement Group
CAIV	Cost as an independent variable
CAS	Cost Accounting Standards
CBU	Cluster Bomb Unit
CCC	Contractor Configuration Control
CDE	Commercial Derivative Engine
CDRL	Contractor Data Requirements List
CEM	Combined Effects Munition
CEP	Circular error probable
CER	Cost estimating relationship
CLU	Command launch unit
CMI	Civil-military integration
CNI	Communications, navigation, identification–friend or foe
COSSI	Commercial Operations and Support Savings Initiative
COTS	Commercial off the shelf
CPAF	Cost plus award fee

CPFF	Cost plus fixed fee
CPIF	Cost plus incentive fee
CPSR	Contractor Purchasing System Review
CSIS	Center for Strategic and International Studies
CSSR	Cost Schedule Status Report
DAPP	Defense Acquisition Pilot Program
DARPA	Defense Advanced Research Projects Agency
DCAA	Defense Contract Audit Agency
DCMA	Defense Contract Management Agency
DCMAO	Defense Contract Management Area Operations
DFAR	Defense Federal Acquisition Regulation
DLA	Defense Logistics Agency
DoD	Department of Defense
DPSC	Defense Personnel Support Center
DRB	Dispute resolution board
DSP	Defense Support Program
EELV	Evolved Expendable Launch Vehicle
EMD	Engineering and manufacturing development
EVMS	Earned Value Management System
FAA	Federal Aviation Administration
FAR	Federal Acquisition Regulation
FARA	Federal Acquisition Reform Act
FASA	Federal Acquisition Streamlining Act of 1994
FDS	Flight Demonstration System

FPIF	Fixed-price incentive fee
FSCATT	Fire Support Combined Arms Tactical Trainer
FY	Fiscal year
GAO	General Accounting Office
GATS	GPS-Aided Targeting System
GBTS	Ground-Based Training System
GEO	Geosynchronus earth orbit
GOTS	Government off the shelf
GPS	Global Positioning System
GTACS	Ground Theater Air Communications System
HAE UAV	High-Altitude Endurance Unpiloted Aerial Vehicle
HEO	Highly elliptical earth orbit
HTI	Hughes Training Inc.
IF	Incentive fee
IIR	Imaging Infrared
ILS	Initial launch services
IMU	Inertial measurement unit
IOT&E	Initial operational test and evaluation
IPPT	Integrated Product and Process Team
IPT	Integrated Product Team
IR&D	Independent research and development
JASSM	Joint Air-to-Surface Standoff Missile
JDAM	Joint Direct Attack Munition
JET	Joint Estimate Team

JPATS	Joint Primary Aircraft Training System
JSF	Joint Strike Fighter
JSOW	Joint Standoff Weapon
KPPs	Key performance parameters
LADS	Low Altitude Demonstration System
LCC	Life cycle cost
LCCV	Low-cost concept validation
LEO	Low-earth orbit
LMTAS	Lockheed Martin Tactical Aircraft Systems
LRIP	Low-rate initial production
MD	Manufacturing development
MDA	Milestone Decision Authority
MDAP	Major Defense Acquisition Program
MIDS	Multifunctional information display system
Mil spec	Military specifications and standards
MMAS	Material Management Accounting System
MRB	Material Review Board
MTVR	Medium Tactical Vehicle Replacement
NDI	Nondevelopmental item
O&S	Operations and support
ORD	Operational Requirements Document
OSD	Office of the Secretary of Defense
OTA	Office of Technology Assessment, "Other Transactions" Authority

OUSD (A&T)	Office of the Under Secretary of Defense, Acquisition and Technology
PA&E	Program Analysis and Evaluation
PAT	Process Action Team
PB	President's budget
PDRR	Program definition and risk reduction
POE	Program Office estimate
PPCC	Procurement price commitment curve
PPV	Past Performance Value
RAM/ILS	Reliability, Availability, Maintainability/Integrated Logistics Support
RDT&E	Research, development, test, and evaluation
RFP	Request for proposal
RHPPC	Radiation-hardened PowerPC
ROM	Rough order of magnitude
SAR	Selected Acquisition Report
SBIRS	Space-Based Infrared System
SFW	Sensor Fuzed Weapon
SLAM-ER	Standoff Land Attack Missile Expanded Response
SOO	Statement of objectives
SOW	Statement of work
SPI	Single-Process Initiative
SPO	System Program Office
TACMS-BAT	Tactical Missile System–Brilliant Anti-Armor Munition

TINA	Truth in Negotiations Act
TRA	Teledyne Ryan Aerospace
TSPR	Total System Performance Responsibility
TSSAM	Tri-Service Standoff Attack Munition
TY	Then year
UAV	Unpiloted air vehicle
UFP	Unit flyaway price
UP	Unit price
URF	Unit recurring flyaway price
USG	United States government
USP	Unit sailaway price
VARTM	Vacuum-assisted resin transfer molding
WCMD	Wind-Corrected Munitions Dispenser

Chapter One
INTRODUCTION AND ACQUISITION REFORM MEASURES TAXONOMY

This report supports a larger RAND project entitled "The Cost of Future Military Aircraft: Historical Cost Estimating Relationships and Cost Reduction Initiatives." Its purpose is to update the technical cost models and cost estimating relationships (CERs) for fixed-wing combat aircraft in light of R&D, manufacturing, organizational, and programmatic advances and reforms that have taken place over the past decade. Given the potential overlap of claimed savings resulting from new (post-1990) aircraft design and manufacturing initiatives (especially for advanced airframe materials), acquisition reform, and lean implementation, the RAND project divided the research effort into the following five areas:[1]

1. New fabrication and assembly processes related to advanced airframe materials;[2]

2. Government changes in acquisition processes or changes in the relationship between the government and Department of Defense (DoD) prime contractors, generally included under the rubric of "acquisition reform";

---

[1] See Appendix A for a listing of all military aircraft initiatives addressed in three of these reports.

[2] Obaid Younossi, Michael Kennedy, and John C. Graser, *Military Airframe Costs: The Effects of Advanced Materials and Manufacturing Processes,* Santa Monica: RAND, MR-1370-AF, 2001.

3. Lean implementation and other initiatives oriented primarily toward processes within a prime airframe manufacturer or relationships between these primes and their suppliers;[3]
4. Technology and process improvements in military avionics development and manufacturing, especially as they relate to the Joint Strike Fighter (JSF);[4] and
5. Technology and process improvements in military aircraft engines (research in progress).

This document reports the results of the research conducted in the second area listed above: acquisition reform (AR). Its purpose is to determine whether published estimates in the literature are sufficiently robust to contribute to the development of adjustment factors for use in predictive cost models that reflect the effects of AR on the costs of developing and producing fixed-wing combat aircraft. The report reviews a wide range of published estimates and projections of claimed savings that may arise from a variety of weapon system AR measures. However, no independent RAND estimates of potential AR cost savings have been generated, nor have any of the published estimates been analyzed. Rather, the existing estimates are grouped into logical categories and compared, and the variations, historical origins, and relative quality of these estimates are discussed. In addition, this document presents the views of numerous prime contractors on potential AR cost savings based on a series of interviews conducted in 1998.[5]

---

[3]Cynthia R. Cook and John C. Graser, *Military Airframe Acquisition Costs: The Effects of Lean Manufacturing Processes,* Santa Monica: RAND, MR-1325-AF, 2001.

[4]Mel Eisman, Jon Grossman, Joel Kvitky, Mark Lorell, Phillip Feldman, Gail Halverson, and Andrea Mejia, *The Cost of Future Military Aircraft Avionics Systems: Cost Estimating Relationships and Cost Reduction Initiatives,* Santa Monica: RAND, limited document, not for public distribution, 2001.

[5]Industry sites visited by RAND include Boeing Military Aircraft and Commercial Aircraft, Seattle, Washington; Boeing McDonnell Military Aircraft and Missile Systems, St. Louis, Missouri; British Aerospace Military Aircraft and Aerostructures, Warton and Samlesbury, United Kingdom; DaimlerChrysler Aerospace Airbus GmbH, Bremen, Germany; DaimlerChrysler Aerospace AG Military Aircraft, Munich, Germany; Lockheed Martin Aeronautical Systems, Marietta, Georgia; Lockheed Martin Skunk Works, Palmdale, California; Lockheed Martin Tactical Aircraft Systems, Fort Worth, Texas; Northrop Grumman Electronic Sensors and Systems Sector, Baltimore, Maryland; Northrop Grumman Integrated Systems and Aerostructures, Air Combat Systems, El Segundo, California; Northrop Grumman Integrated Systems and

The report concludes that on the whole, there is insufficient evidence in the published literature to support the development of precise adjustment factors for AR cost savings that can be used with confidence in technical cost models for military combat aircraft. At the same time, our research suggests that at least in some categories of AR measures, rough order-of-magnitude (ROM) estimates or "rules of thumb" for potential AR cost savings can be developed that may be of use to cost estimators in limited circumstances.

AR became a centerpiece of DoD weapon system procurement policy in the early days of the Clinton administration. Although AR and overhauls of the existing procurement systems had been attempted many times before, the Clinton administration initiatives appeared broader, deeper, and more enduring than many past efforts. Declining post–Cold War defense budgets, growing weapon system costs, and increased technology leadership in the commercial sector prompted then–Secretary of Defense William Perry to emphasize, in his framework documents launching the new push for AR in February 1994, that "change is imperative."[6] Senior DoD officials have maintained and increased their emphasis on AR since Secretary Perry first launched the effort. As Jacques Gansler, the former Under Secretary of Defense for Acquisition and Technology, stressed during testimony before the Senate in 1998, "Acquisition reform is not a slogan. It is a fundamental transformation in our organization, policies, and processes.... Its goals are clear: to do the job better, faster, cheaper."[7]

There is a vast body of literature on AR that covers a wide variety of measures. Innumerable claims have been made for the anticipated or actual cost savings attributable to various aspects of AR. However, many of these claims are difficult to verify and are often inconsistent

---

Aerostructures, Dallas, Texas; Northrop Grumman Integrated Systems and Aerostructures, Hawthorne, California; Raytheon Aircraft, Wichita, Kansas; Raytheon Sensors and Electronic Systems, El Segundo, California; and Scaled Composites, Inc., Mojave, California.

[6]See Secretary of Defense William Perry, *Acquisition Reform—Mandate for Change*, February 1994, and Secretary of Defense William Perry, *Specifications and Standards—A New Way of Doing Business*, June 29, 1994.

[7]Statement by the Honorable Jacques S. Gansler, Under Secretary of Defense, Acquisition and Technology, to the Subcommittee on Acquisition and Technology, Committee on Armed Services, U.S. Senate, March 18, 1998.

or even contradictory. As discussed above, the purpose of this report is to survey the range of estimates available from both published sources and interviews with industry and government sources, to categorize these claims, and to make some initial assessments of the robustness of the estimates.

Since the issuance of former Secretary Perry's initial documents on AR, the DoD, the service executives, Congress, and various other sources have produced many AR measures, initiatives, and policy guidance documents. An array of new AR policies, terminologies, and acronyms has also emerged since Secretary Perry's original pronouncements. Yet the precise definition of these new policies and terms and their relationship to each other have not always been clear or consistent.

Our first task was thus to develop a rational taxonomy of AR measures that would provide a reasonable ordering and coherent linkage between the various measures. Table 1.1 presents our taxonomy of current major AR measures and initiatives, which was developed for the purpose of assigning published cost savings estimates to specific elements. As Table 1.1 indicates, we suggest three principal AR categories: (1) reducing regulatory and oversight burden; (2) commercial-like program structure; and (3) multiyear procurement. Table 1.1 also presents subelements of the second category, which is far more complex and broad than the other two, along with suggestions on how these subelements might be linked to the main category.

Table 1.1 helps clarify and illustrate our interpretation of the key elements of AR and how these elements interrelate. More significantly, it organizes key AR concepts in a way that facilitates the assignment of existing cost savings estimates to specific elements. Many crucial AR measures are included under the second category because we believe that they are inextricably interrelated. The remaining subsections in this report examine all major AR categories as we have defined them and review the data available from published sources and industry interviews regarding both claimed and potential savings.

The tables in the chapters that follow summarize the published AR cost savings estimates and projections in accordance with the cate-

Table 1.1
A Taxonomy of AR Measures

1. Reducing Regulatory and Oversight Compliance Cost Premium
2. Commercial-Like Program Structure
   A. Emphasis on CAIV[a] through the use of:
      (1) Unit price thresholds, unit price targets
      (2) Production price requirement and commitment curves + carrots/sticks in final down-select and in production contract (including warranties, etc.)
      (3) Competition
   B. Enable CAIV through emphasis on:
      (1) Requirements reform
         (a) No "overdesigning"[b]
         (b) Prioritized tradable performance/mission requirements (threshold requirements, etc.)
      (2) Contractor configuration control, design flexibility
      (3) Commercial insertion/dual use,[c] which is made possible by
         (a) Mil spec reform[d]
         (b) Government-industry IPTs[e]
3. Multiyear Procurement[f]

[a]CAIV is an acronym for "cost as an independent variable." The basic concept of CAIV is that it raises rigorous production-unit cost goals to the same priority level as performance and other key system goals during the design and development phases of a weapon system. As such, it is similar to the "must cost" goals that commercial aircraft transport developers and other commercial firms impose on their designers, engineers, and subcontractors when they initiate the development of a new system. More is said on the CAIV concept in subsequent sections of this report.

[b]A more familiar term that could have been used in this context is "gold plating." This term was rejected, however, because some observers associate it with less-than-objective journalistic critiques of the defense acquisition process. The term "overdesigning" as used here means to design into a weapon system capabilities or attributes that may not be worth the extra expense or that are not essential to meeting the mission requirements.

[c]"Commercial insertion" refers to the use of commercial off-the-shelf (COTS) technologies, processes, parts, components, subsystems, and/or systems in weapon systems. The term also refers to the use of "ruggedized" or "militarized" COTS products. "Ruggedization" signifies the special packaging or other hardening of COTS products to permit them to function in harsh military environments. "Dual use" refers to technologies, manufacturing facilities, and products that are known to have or may have both commercial and military applications.

[d]"Mil spec" is an acronym for military specifications and standards.

[e]IPT = Integrated Product Team.

[f]"Multiyear procurement" refers to government authorization for the procurement of specific numbers of production systems beyond the normal single-year government procurement funding cycle. Multiyear procurement requires special congressional approval.

gories and subcategories identified in our taxonomy. Some additional categories are also used as discussed in the body of each chapter. It is important to remember, however, that these estimates vary considerably in both quality and methodology and must therefore be viewed with extreme caution. A detailed reading of the main text of this report is necessary to clarify the many limitations and caveats that must be applied in their use.

Chapter Two discusses what is meant by the DoD regulatory and oversight compliance cost premium. It then presents published estimates of this alleged cost premium grouped under a variety of different categories, each intended to bring together for comparison similar types of estimates. Chapter Three presents detailed case studies of three U.S. Air Force AR pilot programs for the development of new munitions or munitions guidance kits. Chapter Four more briefly reviews a wider variety of AR pilot programs and other types of nonstandard military development efforts. Estimates of savings claimed from multiyear production contracts are presented in Chapter Five.

The final chapter presents in the form of tables all the cost savings estimates from the entire report. Most of the tables have a column labeled "estimate quality" that distinguishes between three kinds of estimates. The highest-quality estimate, labeled "actuals," signifies that the estimate of AR savings was based on actual R&D and production cost data from the specific item under consideration, compared to earlier actuals for the program prior to the imposition of AR. Virtually none of the estimates available during the course of this research effort was of this type.[8]

The second-highest-quality estimate, labeled "forecast," refers primarily to a narrow set of cases in which actual production costs for the specific article are well known but the program is being restructured in a way that is expected to reduce costs. However, no actual costs for the items produced under the restructured program are available for use in the estimate of future cost savings. This applies mainly to estimates of multiyear production contract savings.

---

[8] See the subsequent discussion on actuals.

The third-highest-quality estimates, labeled "analysis," are made in situations where no actual costs are available for the specific item either before or after AR. In such cases, the anticipated pre-AR cost of a specific item that has not yet been developed is compared to the expected cost of that item after the imposition of AR. In other words, neither the actual cost of the item under the old system nor the actual cost of the item after the imposition of acquisition reform is known. This type of estimate is based on rational analysis, past experience, data from analogous military or commercial programs adjusted to the system under consideration, expert opinion, or similar methods.

Almost all the estimates of AR cost savings collected in this report fall into the category we have labeled "analysis"—that is, they are not based on actual cost data for the specific system and the specific program structure in question, either before or after AR. This is another key reason these estimates must be treated with extreme care.

Finally, it is important to reiterate that the information and data cutoff point for this report was December 1999.

Chapter Two

DoD REGULATORY AND OVERSIGHT COMPLIANCE COST PREMIUM

INTRODUCTION

The alleged DoD regulatory and oversight compliance cost premium was one of the first areas examined in detail and targeted for reform by AR advocates. It is therefore appropriate to begin our examination of claimed AR cost savings in this area.

The DoD regulatory and oversight compliance cost premium refers to the additional costs that the DoD is alleged to pay contractors to cover the added cost of complying with the vast array of regulations and requirements imposed on contractors by the government. This cost is claimed to be over and above what the same item would cost were it acquired by a civilian customer in a purely commercial environment.

This chapter includes two main categories of estimates:

1. Direct estimates of the claimed DoD regulatory and oversight compliance cost premium; and
2. Early estimates of overall DoD AR savings.

The direct estimates are discussed in several different groupings based on the period in which they were developed, who developed them, and the quality of the estimates. These categories are discussed more fully below.

Initial DoD estimates of overall AR cost savings are also discussed in this chapter for several reasons. Claimed savings are derived by comparing overall program cost estimates for a large number of Major Defense Acquisition Programs (MDAPs) in various stages of development or production from one budgetary period to another. At the time these savings estimates were made, few new AR pilot programs existed in the databases, and those that did exist were only in the earliest stages of the R&D process. As a result, most of the programs included in these initial AR cost savings studies were benefiting—if they benefited at all—primarily from the effects of reductions in the regulatory and oversight burden. Therefore, these estimates are included under this subsection as part of the determination of the claimed regulatory and oversight cost burden.

DEFINING THE REGULATORY AND OVERSIGHT COMPLIANCE COST PREMIUM

As noted above, an early target for acquisition reformers was the reduction of the government-imposed regulatory and oversight burden—a burden that many observers believed resulted in a significant cost premium for the DoD with little value added and that discouraged commercial firms from doing business with the DoD.

In the late 1980s and early 1990s, a large number of studies conducted both inside and outside the government concluded that the maze of special government laws, regulations, reporting requirements, and policies imposed on contractors doing business with the government had created two serious problems. First, compliance with the laws and regulations by firms, combined with the extra cost of mandated government monitoring and oversight activities, had resulted in a significant cost premium added to items procured by the government. Government regulations often require that companies comply with hundreds of costly and time-consuming reporting rules as well as with similar government-unique accounting and socioeconomic requirements. According to studies conducted at this time, government regulation increased costs to the government by 5 to 50 percent (see Table 2.1).

Second, AR advocates claimed that government-mandated procedures and standards often have not been in conformity with routine business practices in the commercial world—as a result of which many commercial firms have consciously avoided doing business with the DoD. Commercial firms were believed to be unwilling to accept the extra costs and controls on profits or allow government access to the proprietary technical and cost data required to participate in DoD contracts. Those firms that did work on DoD contracts tended either to specialize in military work or to establish separate divisions that were fenced off from their commercial divisions so that government regulations and oversight would not impinge on commercial operations.

The unfortunate result of this situation, according to AR advocates, was twofold. First, the regulatory environment caused the DoD to pay a premium of up to 50 percent more for items it procured than would be the case for similar commercial items. Second, the DoD was denied access to lower-cost, higher-quality commercial products and processes because leading companies refused to do business with it. Many observers therefore regarded the maze of unique government requirements and standards as one of the principal barriers to true integration of the civilian commercial and military industrial bases, often called civil-military integration (CMI).

Thus, most of the DoD regulations and standards identified by early reform studies as driving up contractor costs were also seen as major impediments to civil-military industrial integration and to greater participation of commercial firms in DoD procurement. DoD reform advocates usually viewed the following categories of regulations and standards as the most egregious cost drivers and hence as the greatest barriers to CMI:[1]

- Government access to commercially sensitive product cost and pricing data and certification of cost and pricing data such as those required by the Truth in Negotiations Act (TINA);

- Government-imposed accounting and reporting standards and systems such as Cost Accounting Standards (CAS), the Cost/Schedule Control System Criteria (C/SCSC), and the Material Management Accounting System (MMAS);

---

[1] See Perry, *Acquisition Reform—Mandate for Change,* and Perry, *Specifications and Standards.*

- Audit and oversight requirements such as Defense Contract Management Area Operations (DCMAO) program reviews, Defense Contract Audit Agency (DCAA) audits, and Contractor Purchasing System reviews;
- Complex contract requirements and statements of work (SOWs);
- Socioeconomic and mandatory source requirements; and
- Government ownership and control of technical data.

Table 2.1

Early Subjective Estimates of the DoD Regulatory and Oversight Cost Premium[a]

Study	Date	Estimated DoD Cost Premium/Potential Cost Savings (%)
Honeywell defense acquisition study (20 programs, contractor costs)	1986	13
RAND OSD regulatory cost study (total program costs)	1988	5–10
OTA industrial base study (total DoD acquisition budget)	1989	10–50
CSIS CMI study[b] (cost premium on identical items)	1991	30
Carnegie Commission (total DoD acquisition budget)	1992	40
ADPA cost premium study	1992	30–50

[a]The full titles of these studies are as follows: *Defense Acquisition Improvement Study*, Honeywell, May 1986, G. K. Smith et al., *A Preliminary Perspective on Regulatory Activities and Effects in Weapons Acquisition*, Santa Monica: RAND, R-3578-ACQ, March 1988; Office of Technology Assessment, *Holding the Edge: Maintaining the Defense Technology Base*, Vol. II Appendix, Washington, D.C.: USGPO, April 1989; *Integrating Commercial and Military Technologies for National Security: An Agenda for Change*, Washington, D.C.: Center for Strategic and International Studies, April 1991; *A Radical Reform of the Defense Acquisition System*, Carnegie Commission on Science, Technology and Government, December 1, 1992; and *Doing Business With DoD—The Cost Premium*, American Defense Preparedness Association, 1992.

But exactly how much money could regulatory and oversight reform be expected to save? As shown in Table 2.1, most of the early studies that examined this question used qualitative or theoretical analyses, backed up at best by limited data. Definitions and methodologies often varied significantly from study to study. Not surprisingly, estimates of the size of the cost premium, and thus of the potential savings from regulatory reform, varied dramatically. For example, the Office of Technology Assessment estimated a potential cost savings of 10 to 50 percent in the total DoD acquisition budget, while another study conducted by the American Defense Preparedness Association calculated that product costs for the DoD could be reduced by 30 to 50 percent. Yet at roughly the same time, a more rigorous and considerably more conservative study conducted by RAND suggested that potential savings in terms of total program costs were in the range of only 5 to 10 percent.

Although these and other studies sometimes lacked precision or analytical rigor and offered widely different assessments of potential savings, they nonetheless had significant impact. Many influential members of Congress, as well as senior DoD officials and defense analysts, accepted the studies' basic premise that the DoD regulatory burden (1) imposed a significant cost compliance premium on DoD procurement, and (2) prevented participation of the commercial sector in weapon system development, thereby further raising costs and lowering quality.

In response to these concerns, Congress passed Section 800 of the National Defense Authorization Act of 1990, which required that the DoD establish a panel of experts from government, industry, and academia to evaluate changes to DoD acquisition regulations. The "Section 800 Panel" recommended eliminating or changing about one-half of the 600 statutes it identified that affect DoD acquisition. The panel's findings were submitted to Congress in January 1993 for legislative action.[2]

---

[2]See Statement of the Under Secretary of Defense for Acquisition and Technology, Honorable Paul G. Kaminski, Before the Acquisition and Technology Subcommittee of the Senate Committee on Armed Services on Defense Acquisition Reform, Committee on Armed Services, U.S. Senate, March 19, 1997.

Based on the Section 800 Panel findings and the work of former Vice President Gore's National Performance Review, the DoD developed an AR strategy that then–Secretary of Defense William Perry presented to Congress in February 1994 in his document titled *Acquisition Reform—Mandate for Change*. This document called for a much more flexible, commercial-like acquisition approach that emphasized the importance of CMI and the acquisition of commercial products, technologies, and processes. A Deputy Under Secretary of Defense for Acquisition Reform was also appointed and a Process Action Team (PAT) was formed to examine the reform of military specifications and standards (mil specs) and reductions in government regulation and oversight of contractors. The PAT's report called for replacing mil specs with performance specifications or existing commercial standards wherever practical. Secretary Perry ordered the implementation of these recommendations in June 1994.[3]

In many instances, reducing the regulatory burden and promoting CMI required legislative action by Congress. Accordingly, many of the DoD's AR concepts were incorporated into the Federal Acquisition Streamlining Act (FASA) of 1994. FASA greatly simplified DoD procedures for purchasing relatively low-cost, low-risk commercial products and services. The act also changed the definitions of commercial and nondevelopmental items and exempted these items from many DoD acquisition regulations and requirements. Finally, FASA authorized the establishment of Defense Acquisition Pilot Programs (DAPPs) to test out more radical modes of AR.[4]

---

[3]See Perry, *Specifications & Standards*.

[4]More is said on DAPPs below. Signed on October 13, 1994, FASA sought to make government acquisition of commercial goods and services easier. Toward this end, it (1) expanded the definition of commercial items; (2) automatically exempted the purchase of commercial items from more than 30 government-unique statutes; (3) removed the requirement for cost and price data on commercial contracts; (4) raised the threshold for the application of TINA to $500,000; and (5) expanded the information provided to all competitors after contract award to reduce formal protests. The Federal Acquisition Reform Act of 1996 (FARA) made additional changes in efforts to promote even greater government access to the commercial marketplace by further simplifying procedures for purchasing certain categories of commercial items. See Office of the Under Secretary of Defense, Acquisition and Technology, *Overcoming Barriers to the Use of Commercial Integrated Circuit Technology in Defense Systems*, October 1996, Appendix B.

THE COOPERS & LYBRAND STUDY

Yet much of the regulatory burden imposed on contractors was not based directly on legislation, arising instead from unique DoD rules and requirements. In early 1994, Secretary Perry tasked a private consulting firm, Coopers & Lybrand (C&L), to undertake a detailed analysis of the costs of industry compliance with these regulations so that the DoD could target the most important cost drivers in its quest for acquisition regulatory reform. C&L then conducted an extensive data collection effort at ten defense contractor sites[5] focusing on 130 DoD regulations and standards that were identified by the Section 800 Panel and by other studies as being major cost drivers and impediments to CMI.

C&L explicitly evaluated only the direct costs of compliance with DoD regulations. By and large, these costs should be considered overhead costs associated with data collection, report and proposal preparation, inspection, auditing, and the like. C&L did not include the potential additional cost savings of using commercial standards, processes, technologies, parts, and components. In addition, C&L applied a methodology based on "activity-based costing" (ABC) and examined only the portion of the contract cost that was value-added by the contractors under investigation. Using these assumptions and this methodology, C&L concluded that on average, the DoD paid a regulatory cost premium of approximately 18 percent.[6]

C&L's findings, reported to the DoD in December 1994, proved to be a highly influential and often-cited document. First and perhaps most significantly, it was widely considered to be the first truly objective assessment of the DoD regulatory cost premium—i.e., the first to be based on a detailed assessment of an extensive and sys-

---

[5]These sites were Allison Transmission (a subsidiary of General Motors), Beech Aircraft (a division of Raytheon), Boeing Defense and Space Group, Rockwell Collins Avionics and Communications Division, Hughes Space and Communications Company (a subsidiary of General Motors), Motorola Government Systems Technology Group, Oshkosh Truck-Chassis Division, the Timken Company, Teledyne Ryan TCAE Turbine Engines, and Texas Instruments Defense Systems and Electronics Group. Some of these companies have since merged or been acquired by other entities.

[6]See Coopers & Lybrand/TASC, *The DoD Regulatory Cost Premium: A Quantitative Assessment*, December 1994.

tematically collected database. Second, although it tended to be on the lower end of the spectrum of earlier studies of the DoD cost premium as discussed above, the C&L study seemed to show that significant savings were still potentially achievable through reductions in DoD regulation and oversight. This is particularly true because C&L explicitly excluded any savings that might result from greater use of commercial technologies, processes, and parts. In other words, the estimated 18 percent DoD cost premium was due solely to compliance costs with DoD-unique regulations. Third, the C&L study suggested that large savings could be gained by eliminating or reforming only a handful of regulations. It found that the top three cost drivers accounted for more than 20 percent of the total average DoD regulatory cost premium, while the top ten accounted for about half, as shown in Table 2.2. Finally, the study concluded that the top 24 cost drivers accounted for 75 percent of the DoD cost premium.

A brief description of the top ten C&L cost drivers follows.

- As indicated in Table 2.2, **MIL-Q-9858A** was identified as the most significant cost driver. This is an umbrella standard that establishes a basic framework for implementing quality control measures in all areas of contract performance. C&L found that it required excessive documentation and reporting as well as unnecessary and repetitive testing compared to widely accepted commercial quality control standards such as ISO-9000.

Table 2.2

DoD Regulatory Compliance Cost Premium: C&L Top Ten Cost Drivers

Cost Driver	Percentage of Total Cost Premium
1. MIL-Q-9858A	10.0
2. TINA	7.5
3. C/SCS	5.1
4. Configuration management	4.9
5. Contract requirements/SOW	4.3
6. DCAA/DCMAO interface	3.9
7. CAS	3.8
8. MMAS	3.4
9. Engineering drawings	3.3
10. USG property administration	2.7

- **TINA** was found to require highly detailed certified cost and pricing data in contract proposals. The data must be generated and supplied at least three times during the contract process. Compliance costs were found to be high in part because of government auditing requirements. Criminal penalties can be imposed on company officials if irregularities are found in cost and pricing data which have been submitted in proposals.

- The **C/SCS** category in Table 2.2 includes the DODI 5000.2 Cost/Schedule Status Report (CSSR), Defense Federal Acquisition Regulation (DFARS) 252.234-70001 Cost/Schedule Control System Criteria, and MIL-STD-881 Contract Work Breakdown Structure. Compliance costs for C/SCS were found to be high because of detailed and burdensome reporting and tracking requirements that are not routine in usual commercial practice.

- **Configuration management** is based on MIL-STD-973. Although tracking and documenting engineering changes are considered to be crucial tasks, industry argues that MIL-STD-973 requires excessive documentation and is too complex.

- **Contract requirements** and **SOW** issues refer to the extreme complexity of DoD contracts compared to commercial contracts and to the imposition of process requirements.

- Some contractors perceive DoD on-site **DCAA** and **DCMAO** representatives as engaged in unnecessary and costly intrusions into their normal business and manufacturing activities.

- **Cost accounting standards** impose government-unique cost accounting requirements that vary from standard commercial practice.

- The **MMAS** requires that contractors collect extensive cost data by contract on materials.

- DoD requirements for **engineering drawings** vary considerably from standard commercial practice.

- Finally, **government property administration** rules impose complex and costly bookkeeping requirements on contractors who use government-owned equipment.

One reason the C&L study was so influential is that its broad conclusion appeared to be confirmed by other studies. The Principal Deputy for Acquisition, U.S. Army Material Command (AMC), for example, directed NORCOM, a private consulting firm, to undertake a study similar to that of C&L with the goal of determining the cost of Army contractors' compliance with DoD regulations. NORCOM's AMC study applied activity-based cost analysis to data collected from six U.S. Army contractors, most of whom specialized in military-unique items such as machine guns. In its final report dated May 1994, NORCOM estimated that the weighted-average DoD regulatory cost premium amounted to 27 percent.[7] This number is close to C&L's estimate of a 22 percent regulatory compliance cost for companies that produce military unique items for the DoD. NORCOM's top four cost drivers were also similar to those of C&L, even though the AMC study used either broader categories or categories that were not exactly comparable in other ways. Thus, in the AMC study, government quality systems, auditing and accounting requirements, and contracting and pricing regulations accounted for more than 50 percent of the DoD regulatory compliance cost premium.

EARLY RESULTS OF DoD INITIATIVES TO REDUCE THE DoD REGULATORY AND OVERSIGHT BURDEN

In response to the C&L study and to similar studies such as the NORCOM effort, the DoD established the Regulatory Cost Premium Working Group to investigate reforming or eliminating the top C&L cost drivers. In September 1994, the DoD also established the DoD Reducing Oversight Costs Reinvention Laboratory. Ten contractor sites participated along with government officials from the Office of the Secretary of Defense (OSD), the Defense Contract Management Command (DCMC), and DCAA.[8] The participants conducted exten-

---

[7]See *Activity-Based Cost Analysis of Cost of DoD Requirements and Cost of Capacity: Executive Summary*, NORCOM, May 1994. The average cost premium was derived by applying a weighting to the results from each of the six firms based on total sales revenue and total DoD business.

[8]The ten sites were at Boeing, Seattle, Washington; Northrop Grumman, Hawthorne, California; Hughes Missile Systems, Tucson, Arizona; Lockheed Martin, Fort Worth, Texas; Loral Vought, Grand Prairie, Texas; Texas Instruments, Dallas, Texas; McDonnell Douglas, St. Louis, Missouri; Magnavox, Fort Wayne, Indiana; Lockheed Martin, Moorestown, New Jersey; and Raytheon, Bedford, Massachusetts.

sive cost/benefit analyses on reducing oversight and regulatory requirements and reported their results in a manner that was based on the C&L categories.[9]

By mid-1995, the results of these efforts began to be reported back to the high-level DoD leadership. These results, however, were somewhat less encouraging than initial expectations. The Regulatory Cost Premium Working Group focused its primary efforts on actions to mitigate the effects of the top 24 cost drivers, identified by the C&L study, which accounted for 75 percent of the DoD regulatory cost premium. According to the C&L study, these 24 cost drivers led to an average DoD cost premium of 13.4 percent. The Working Group eventually concluded, however, that the DoD could reasonably expect to achieve cost savings of only 46 percent of the cost premium claimed by C&L for the top 24 cost drivers, for a total estimated average cost savings of only 6.3 percent. This was because the Working Group concluded that retention of some elements of the regulations identified as the top 24 cost drivers was necessary for maintaining public trust and pursuing beneficial aspects of oversight. The Working Group also pointed out that even the potential of 6.3 percent savings was probably optimistic because it did not reflect the implementation costs of reform and the substitution of new measures where necessary.[10]

Meanwhile, an extensive General Accounting Office (GAO) study of the Reducing Oversight Costs Reinvention Laboratory effort also concluded that the C&L estimates of the potential savings from mitigating the DoD regulatory and oversight burden were probably optimistic. GAO reported that five of the ten participants in the Reinvention Laboratory had prepared their own estimates of the cost impact at their sites of the top ten C&L cost drivers. These estimates ranged from 1.2 to 6.1 percent compared to the C&L estimate of 8.5

---

[9]See U.S. General Accounting Office, *Acquisition Reform: Efforts to Reduce the Cost to Manage and Oversee DoD Contracts,* GAO/NSIAD-96-106, April 1996.

[10]See Office of the Under Secretary of Defense, Acquisition and Technology, Acquisition Reform Senior Steering Group, DoD Regulatory Cost Premium Working Group, *Updated Compendium of Office of Primary Responsibility (OPR) Reports,* June 1996; and Office of the Under Secretary of Defense (Acquisition and Technology), Acquisition Reform Senior Steering Group, DoD Regulatory Cost Premium Working Group, *Compendium of Office of Primary Responsibility (OPR) Reports,* June 30, 1995.

percent. In addition, participants experienced little success in addressing nine of the top ten cost drivers. Almost all projected savings came from converting from the mil spec quality control system (MIL-Q-9858A) to commercial or international standards.[11]

Based on the lessons learned from the Reducing Oversight Costs Reinvention Laboratory effort, the DoD developed the Single-Process Initiative (SPI), which was launched by Secretary Perry with a widely circulated memo in December 1995.[12] SPI is intended to reduce the DoD cost premium and to eliminate many of the regulatory barriers identified by the C&L study as major DoD cost premium drivers by promoting block changes to the manufacturing and management requirements of all existing contracts on a facility-wide basis. Its goal is to eliminate multiple, duplicative, and government-unique management and manufacturing processes at defense contractor installations—processes required by numerous existing defense contracts and DoD regulations—and to replace them with commercial or internationally accepted management and manufacturing processes that are standardized across all contracts at the same facility.[13]

Since the launching of the initiative, SPI has clearly achieved many successes. By October 1998, 300 contractor facilities, including representatives of more than 80 percent of the top 200 DoD contractors, had participated in SPI. A total of more than 1000 block change modifications had been accepted out of nearly 1500 that had been proposed.[14]

DCMC closely tracked the progress of SPI and collected considerable data on SPI cost savings and cost avoidance, which were certified by DCAA. Since the data are collected primarily by facility and by broad category of block change, however, it is difficult to estimate the overall regulatory cost premium savings that have accrued to DoD

---

[11] General Accounting Office, *Acquisition Reform: DoD Faces Challenges in Reducing Oversight Costs*, GAO/NSIAD-97-48, January 1997.

[12] See Secretary of Defense William J. Perry, *Common Systems/ISO-9000/Expedited Block Changes*, December 6, 1995.

[13] See Office of the Deputy Under Secretary of Defense (Acquisition Reform), *Single Process Initiative, Acquisition Reform Acceleration Day Stand-Down*, 1996.

[14] Defense Contract Management Command, *Single Process Initiative Implementation Summary*, October 9, 1998.

for procurement.[15] Many of the estimated savings are clearly due to the adoption of "best business practices" or aspects of lean manufacturing and thus should not be directly attributed to the removal of the DoD regulatory compliance cost burden. In late 1998, DCMC data showed that SPI had resulted in some $30.3 million in direct cost savings to the DoD ("negotiated consideration") and in roughly $475.2 million in "extended cost avoidance," defined as estimated cost savings over the lifetime of all contracts affected by the block changes.

As a crude measure of the relative scale of these savings, a comparison can be made to the overall DoD procurement and RDT&E budgets. As a percentage of the FY98 procurement and RDT&E budgets, the total direct cost savings from SPI amount to only 0.03 to 0.04 percent. The total lifetime cost avoidance to date from SPI stood at some 0.5 percent of the DoD's FY98 procurement and RDT&E budget. Interestingly, one contractor who explicitly attempted to calculate the savings from SPI on a specific program also came up with a savings of 0.5 percent. While these are not particularly precise or revealing comparisons, they do not contradict the consensus view that began to emerge in 1995 that the C&L estimates for potential DoD cost premium savings were too high.

For some data from specific pilot programs, at least some of the factors identified by the C&L study can be examined in isolation for their contribution to total cost savings. These data, although limited, are in the same general range as the final estimates of the Reducing Oversight Cost Reinvention Laboratory and the DoD Regulatory Cost Premium Working Group. Two examples are discussed below.

The breakouts of AR categories for the Wind-Corrected Munitions Dispenser (WCMD) and the Fire Support Combined Arms Tactical Trainer (FSCATT), which are shown in later sections of this report and which are attributed to the category of regulatory and oversight burden, show numbers similar to the other sources discussed above. For WCMD, 3.5 percent of the costs of a traditional R&D program were saved (3.2 percent for production) by reducing an identifiable

---

[15]There are 40 reporting categories, few of which clearly correlate with C&L cost drivers or specific programs. Examples include Quality-Calibration, Manufacturing-Management, Quality-Supplier, and Logistics-Packaging.

factor attributable to the category of regulatory burden as defined here (see Table 7.2).[16] In the case of FSCATT, roughly 2 percent of the likely traditional program costs were expected to be saved by reducing the regulatory burden. Finally, a draft study conducted by the U.S. Air Force Material Command (AFMC) showed that the savings from a major component usually included in the regulatory burden category resulted in a 2.3 percent savings for the total production contract for the B-2 Air Vehicle #1 Upgrade program.[17]

EARLY NON-DoD ASSESSMENTS OF OVERALL DoD AR SAVINGS

Finally, in this subsection we also quickly review initial attempts during the Clinton administration to estimate overall savings on all programs from AR. The reason these estimates are included here is that the vast majority of programs included in the databases that supported these early estimates had been under way for some time as traditional programs. Little time had passed since the beginning of the new phase of AR to permit radical AR pilot programs to truly get under way. Thus, it is not unreasonable to assume that the cost savings—if any—that are identified in these studies are due largely to a reduction in the regulatory and oversight compliance cost premium, which was the initial AR target of opportunity for the DoD.

An important early goal of AR advocates was to collect data demonstrating the cost benefits of AR. In 1995, officials in the Office of the Under Secretary of Defense, Acquisition and Technology (OUSD [A&T]) became concerned with the lack of consistent methodologies and measures of merit in the reporting of AR savings. In March 1996, OSD officials thus tasked the services and the Defense Logistics Agency (DLA) with providing uniform AR cost savings data. The methodology eventually adopted depended on comparing program budgets in the 1997 President's budget (97PB) to the 1995 President's

---

[16]The WCMD assessment attributes these savings to the elimination of Contractor Data Requirements Lists (CDRLs).

[17]*Acquisition Reform Cost Savings and Cost Avoidance: A Compilation of Cost Savings and Cost Avoidance Resulting from Implementing Acquisition Reform Initiatives*, AFMC draft report, December 19, 1996. This B-2 upgrade program benefited from a large reduction in the number of CDRLs that had to be prepared for the government.

budget (95PB). In addition, the services were required to provide estimates of program savings out to fiscal year 2002 (FY02). This exercise resulted in a total DoD savings estimate of $29 billion.

At least three outside studies examined these or similar data for the purpose of independent analysis, and all three raised serious doubts about the level of savings claimed by the services. However, at least two of the studies estimated savings that are well within the range of savings estimated by the Reducing Oversight Costs Reinvention Laboratory and the Regulatory Cost Premium Working Group.

The first study originated when OUSD (A&T) requested that RAND assist OSD in analyzing the service data. Specifically, OSD asked RAND to help standardize the data and conduct a preliminary analysis. Toward this goal, RAND assessed 70 MDAPs using Selected Acquisition Report (SAR) data. The RAND study concluded that the total estimated savings amounted to $22 billion but that the bulk of the reported cost reductions represented future cost avoidance expected in FY02 and beyond. The study estimated that the actual AR cost savings for FY95 through FY01 amounted to some 3.5 percent of total program budgets. If cost avoidance for this period is included, the total rises to 4.4 percent.[18]

Another study, carried out mostly in 1996, used a similar approach and yielded similar results. This study, conducted by a Coast Guard officer and published as a master's thesis in the Management of Technology program at MIT, conducted a detailed examination of 23 MDAPs reporting significant AR success and compared estimated cost savings to actual program budget data. The study concluded that average cost savings plus cost avoidance equaled 4.3 percent—almost the same figure produced by the RAND study.[19]

Finally, a GAO study published in October 1997 analyzed the service reports of AR savings using a methodology similar to the 1996 RAND study. The GAO study also concluded that the service-reported AR

---

[18]John Schank, Kathi Webb, Eugene Bryton, and Jerry Sollinger, "Analysis of Service-Reported Acquisition Reform Reductions: An Annotated Briefing," unpublished research, September 1996.

[19]Lieutenant Commander Michael H. Anderson, *A Study of the Federal Government's Experiences with Commercial Procurement Practices in Major Defense Acquisitions*, Cambridge, MA: Massachusetts Institute of Technology, June 1997.

savings were overstated. It noted that only about 25 percent of the $29 billion in reported savings represented reductions from approved budgets and took place between FY95 and FY02. GAO also evaluated and compared SAR data from 1993 and 1995 for 33 weapon programs that accounted for more than 60 percent of the reported AR savings, and it found that more than two-thirds of these programs actually experienced cost growth after adjustments were made for inflation and quantity changes. The average for all 33 programs was a cost growth of 2 percent. GAO concluded that AR savings did not necessarily lead to reductions in overall program costs because the cost savings were offset by cost growth elsewhere or by reinvestments in the programs.[20]

It is important to emphasize, however, that GAO did not dispute the claim that AR produces real cost savings. Rather, it argued that those savings were (1) overstated by the services, and (2) often offset by other factors, resulting in no reduction in overall program costs. Yet even the GAO analysis showed that 10 of the 33 programs evaluated showed real overall program cost savings ranging from 0.3 to 19 percent, with an average cost decrease of 4 percent. Thus, GAO's 4 percent savings for the ten successful programs is very close to the RAND number of 4.4 percent and to the MIT number of 4.3 percent.

ADDITIONAL OBSERVATIONS FROM INDUSTRY AND GOVERNMENT INTERVIEWS

In 1998, RAND researchers interviewed a wide range of managers at most of the main facilities where Boeing, Lockheed Martin, and Raytheon design and develop fixed-wing military aircraft. One area of discussion covered their experiences with AR and its potential cost savings. Managers representing some of these companies' commercial aircraft divisions were interviewed. In addition, some major military avionics companies were interviewed. Finally, RAND interviewed various government officials involved in AR issues, in-

---

[20]U.S. General Accounting Office, *Acquisition Reform: Effect On Weapon System Funding*, GAO/NSIAD-98-31, October 1997.

cluding representatives of the DoD, some of its agencies, and the three services.[21]

This subsection reports interviewees' views on the DoD regulatory and oversight compliance cost premium and on the initial estimates of overall DoD AR cost savings as discussed above. A later subsection provides the views of industry and government officials interviewed on other aspects of AR. To avoid disclosing proprietary information, specific companies and programs are usually not mentioned in the body of this subsection.

Almost all contractors interviewed strongly agreed that the C&L study and similar studies were correct in concluding that the traditional DoD regulatory and oversight regimen imposes a significant cost premium on DoD purchases. However, nearly all contractors believed that the potential savings were exaggerated, and nearly all had concerns about the implementation of regulatory reform. In summary, contractors made the following points:

- In principle, savings can be realized from reducing the DoD regulatory and oversight burden, but C&L's estimates of potential savings are too optimistic.

- It is difficult to separate out overhead savings due to AR from those due to other factors. Any AR overhead savings are probably overwhelmed by the decline in the business base.

---

[21]Foreign contractors were also interviewed. Industry sites visited include Boeing Military Aircraft and Commercial Aircraft, Seattle, Washington; Boeing McDonnell Military Aircraft and Missile Systems, St. Louis, Missouri; British Aerospace Military Aircraft and Aerostructures, Warton and Samlesbury, United Kingdom; DaimlerChrysler Aerospace Airbus GmbH, Bremen, Germany; DaimlerChrysler Aerospace AG Military Aircraft, Munich, Germany; Lockheed Martin Aeronautical Systems, Marietta, Georgia; Lockheed Martin Skunk Works, Palmdale, California; Lockheed Martin Tactical Aircraft Systems, Fort Worth, Texas; Northrop Grumman Electronic Sensors and Systems Sector, Baltimore, Maryland; Northrop Grumman Integrated Systems and Aerostructures, Air Combat Systems, El Segundo, California; Northrop Grumman Integrated Systems and Aerostructures, Dallas, Texas; Northrop Grumman Integrated Systems and Aerostructures, Hawthorne, California; Raytheon Aircraft, Wichita, Kansas; Raytheon Sensors and Electronic Systems, El Segundo, California; and Scaled Composites, Inc., Mojave, California.

- Selective implementation of regulatory relief through waivers rather than through a permanent end to the regulations significantly reduces potential cost savings.
- Many barriers still exist to implementing the recommendations of the C&L report and other similar studies.
- Some field-level government officials are resisting full implementation of regulatory and oversight reform measures.
- Most SPI savings are really future cost avoidance. SPI savings due to regulatory relief are difficult to identify and quantify.
- The initial estimates of overall AR savings made in the 1995–1997 time frame are difficult to verify and are probably not very reliable.

Most contractors estimated that the potential savings from reform of C&L's top ten cost drivers are on the order of 4 to 6 percent, although some firms place the potential figure as high as 15 percent. Because of the implementation problems discussed below, one contractor insisted that the savings so far from trying to reform C&L's top ten cost drivers were 1 to 2 percent at most, although this contractor accepted the estimated savings potential of 4 to 6 percent with full implementation.

Most contractors agreed that, all things remaining equal, regulatory and oversight reform savings should be reflected primarily in reductions in forward-pricing overhead rates. However, some contractors noted that it is difficult to separate the effects of downsizing and mergers from regulatory reform as factors causing a reduction in contracting and other overhead personnel. Furthermore, many contractors stressed that the decline in the overall business base overwhelms any overhead reductions due to regulatory reform. None of these firms had systematically collected actual data to support AR savings claims.

Firms singled out TINA as the most onerous regulatory burden listed in the C&L top ten. Other C&L cost drivers often mentioned included C/SCS, MMAS, CAS, and MIL-Q-9858A and 1520 (corrective action procedures). The main negative cost effect of TINA, according to many industry representatives, is in the area of proposal preparation and implementation. Managers claimed that TINA-compliant proposals require much more paperwork, fact-finding documentation, audits, and background information than non-TINA proposals. Several contractors noted that many TINA waivers had been granted, resulting in reduced numbers of people involved in contracts, pricing, and other aspects of proposal preparation for specific projects. Officials at one firm claimed that TINA-compliant proposals required twice the personnel that would otherwise be needed. They said this was clearly demonstrated when they prepared a commercial contract proposal for a foreign customer for a weapon system that had previously been sold to the U.S. government.

TINA waivers were said to reduce the cycle time for proposal preparation by 50 percent. Another contractor stated that because of TINA and other regulatory waivers, the company had saved roughly 10 percent of the cost of preparing one major weapon system proposal. According to a third contractor, the number of company personnel dealing with government questions on overhead rates had been reduced by three-quarters as a result of AR.

Many contractors observed, however, that the selective application of TINA and other regulatory waivers to specific programs undermined the realization of much of the potential savings. Since other programs at the same facility still required certified cost and pricing data, many of the specialized pricing and contracting personnel still had to be retained. It was also noted that TINA waivers vary from service to service. Programs were still required to provide some cost and pricing data to the government as a substitute for TINA, but since the data requirements of the services vary, even more work was generated for the contracting personnel. As a result, according to one manager, the actual reduction in contracting and pricing personnel at his facility was "nonexistent or at least very small, definitely well under 10 percent." Thus, at one facility, an AR pilot program was being charged the same overhead rate as all the other traditional programs under way at the same facility.

Some contractors focused on other cost drivers identified in the top ten C&L list and on the problems associated with reforming them. One noted that failure to fully address the problems of U.S. government property management regulations was the "biggest failure" of regulatory reform. Many complained about perceived problems associated with reforming the C/SCS. One contractor said that the government had made no effort to encourage replacement of the costly and cumbersome C/SCS. This contractor also asserted that replacement of this system with another would be costly and disruptive. Another contractor explained that the government required the implementation of Earned Value Management System (EVMS) reporting as a substitute for C/SCS. EVMS required the collection of cost data on at least one level below what this contractor would have done on a typical commercial program. The contractor claimed that EVMS cost millions of dollars to set up and required more than three times as many people than would be needed for the same type of work on a commercial program.

Some firms claimed to have positive, constructive relationships with on-site government personnel and government program officials, while others perceived a more adversarial and intrusive relationship. At some locations, DCMC officials had clearly reduced intrusive inspection and other activities criticized by contractors. One DCMC official noted that in the past all Material Review Board (MRB) actions had to be reviewed and approved but that DCMC had changed to random checks as part of an integrated product team (IPT). This, it was claimed, had reduced cycle time and was less disruptive. At another site, an official noted that processes that were DCMC-inspected had been reduced by two-thirds.

On the negative side, there were also many complaints. As one manager put it, "lower-level government people haven't quite gotten the word yet about acquisition reform." Another manager concluded:

> We can buy commercial parts, but we can't do it commercially.... DCMC imposes enormous documentation requirements, even for piece parts. Acquisition reform means going from 120 pages of documentation down to 50 pages, instead of down to the one page that would be required on a commercial program.

One service official's view that "the organizational bureaucracy resists acquisition reform" was widely held by other service AR offices.

Almost all contractors and government officials interviewed were unanimous in their praise of SPI. It was called "a real success story that permits contractors to standardize for all government customers." Nonetheless, most admitted that the actual scale of real savings from SPI was relatively modest compared to the overall DoD procurement budget. Most contractors and government officials agreed with the official DCMC data that showed that most SPI cost savings were really future cost avoidance. In principle, according to one contractor, SPI savings should be reflected in overhead rates, distributed direct rates, and task direct labor hours. Some OSD officials pointed out that only cost-plus contracts would benefit, not fixed-price contracts.

Many argued that it was usually difficult to match specific savings to specific SPI measures. Government officials noted that normal cost accounting procedures do not track process costs. Thus, it would be difficult to verify and audit specific SPI cost savings. They claimed that DCMC data on SPI savings are mainly ROM numbers. As one OSD official claimed, "You can't make any global conclusions about DCMC/DCAA data. You can't apply it to specific programs or what it means for overall acquisition reform savings."

Finally, government officials involved in the initial collection and analysis of the data on overall AR cost savings as discussed above argued that these early estimates are difficult to verify. OSD officials were highly skeptical about the reliability of the early (1995–1996) government estimates. One official said that "your guess is as good as ours" regarding the true level of savings from AR. Service AR offices were generally equally skeptical. One service official complained that it was "extremely difficult to get accurate acquisition reform savings numbers." Individual programs and program managers supply estimates, but there was not always uniformity in methodology and analytical approach. An OSD official's lament that "we really don't know what the acquisition reform savings are" was reinforced by a service official's admission that AR savings numbers were "pulled out of the air."

OVERSIGHT COMPLIANCE COST PREMIUM SUMMARY AND COST SAVINGS OBSERVATIONS

Based on the data available, it is impossible to provide a precise estimate of the cost savings that can be expected to accrue from regulatory and oversight reform, much less from the specific elements that go into it. It seems clear that there is some level of cost premium paid by the DoD by virtue of the regulatory and oversight burden imposed on contractors. The available evidence suggests, however, that a significant reduction in the regulatory and oversight cost premium is likely to result at best in only relatively modest savings. Nearly all the credible direct and indirect estimates seem to fall within a range of about 6 percent savings or less. Therefore, we believe that a plausible rule-of-thumb estimate of the potential program savings from regulatory and oversight reform is 1 to 6 percent, with an average of 3 to 4 percent. In other words, if one is using a pre-AR program (prior to 1994) as an estimating analogy for a similar new program, cost reductions at the program acquisition level of 3 to 4 percent can reasonably be attributed to reductions in the regulatory and oversight burden.

However, if the cost analysis is developed using prior program direct or indirect labor hours, most of the AR savings from reductions in regulatory and oversight burdens should already be reflected in the negotiated Forward Pricing Rate Agreements (wrap rates), so no further adjustment would be warranted in the rates themselves. This is because most regulatory burden cost savings are in the area of indirect costs and should thus show up in reduced overhead rates. Because AR has been in existence since 1995, most of the realizable reductions in regulatory and oversight burdens should already have been calculated between the contractor and the Defense Contract Management Agency (DCMA).

AR reductions between suppliers and the prime may have to be assessed separately. Factors such as regulatory flow-down and the cost effects of strategic supplier relationships need to be taken into account. Although AR focuses mainly on interactions between the government and the primes, there may be areas between primes, subcontractors, and suppliers that result in further savings.

Chapter Three

COMMERCIAL-LIKE PROGRAM STRUCTURE

INTRODUCTION

Many advocates argue that the greatest potential savings from AR during the R&D as well as production phases may arise from a large group of interrelated measures and reforms that, when applied together as a package, help radically transform traditional government weapon system R&D programs into more commercial-like programs. The U.S. government has already begun testing many comprehensive packages of commercial-like approaches to military acquisition in a variety of innovative pilot and demonstration programs. We examine a sample of ten official pilot and demonstration programs, divided into four categories determined either by weapon system type or by pilot program category.

The first category is made up of case studies of three guided-munition AR pilot programs led by the U.S. Air Force: Joint Direct Attack Munition (JDAM), Wind-Corrected Munitions Dispenser (WCMD), and Joint Air-to-Surface Standoff Missile (JASSM). These programs are examined in greater detail than the others because of their high profiles and the wide variety of information available on them.[1]

Second, we examine two U.S. Air Force space system AR pilot programs: Evolved Expendable Launch Vehicle (EELV) and Space-Based

---

[1] These three case studies are also reviewed in Mark Lorell, Julia Lowell, Michael Kennedy, and Hugh Levaux, *Cheaper, Faster, Better? Commercial Approaches to Weapons Acquisition*, MR-1147-AF, Santa Monica: RAND, 2000.

Infrared System (SBIRS). The third category reviews two additional DAPPs: Fire Support Combined Arms Tactical Trainer (FSCATT) and Joint Primary Aircraft Training System (JPATS).

In the final category, we examine three Defense Advanced Research Projects Agency (DARPA) Section 845 "Other Transactions" programs for major weapon system platforms: Tier III– DarkStar, Tier II+ Global Hawk, and the Arsenal Ship.

The AR pilot programs examined here generally employ a wide variety of specific reform measures intended to ensure the achievement of the anticipated theoretical benefits of CMI. Many of these measures are drawn from or attempt to replicate conditions in commercial markets. All are intended to promote the use of commercial parts and technologies and to encourage the participation of commercial firms in order to reduce costs and increase quality. At the same time, they are designed to encourage the types of market-driven safeguards that usually ensure competitive pricing and high quality in normal commercial markets.

Table 3.1 repeats the basic principles and interrelationships in a commercial-like weapon system program structure based on CAIV as we defined it in Table 1.1. Before the ten pilot program costs are examined in detail, some additional discussion of the factors in Table 3.1 is necessary.

Probably the single most important element for carrying out this transformation to a commercial-like weapon system R&D approach is the concept of cost as an independent variable (CAIV). AR advocates often implicitly link CAIV to requirements reform and to a conscious policy of commercial parts and technology insertion, as we have done here. CAIV is a popular DoD reform concept whose definition varies somewhat depending on the source. In December 1995, a Defense Manufacturing Council working group produced a report that advocated a strategy of aggressive cost objectives for defense systems.[2] In November 1995, Noel Longuemare, the Principal

---

[2]See Dr. Benjamin C. Rush, "Cost as an Independent Variable: Concepts and Risks," *Acquisition Review Quarterly*, Spring 1997. Raymond W. Reig provides a detailed

Table 3.1

Elements of a Commercial-Like Program Structure

A. Emphasis on CAIV through the use of:
   (1) Unit price thresholds, unit price targets
   (2) Production price requirement and commitment curves + carrots/sticks in final down-select and in production contract (including warranties, etc.)
   (3) Competition

B. Enable CAIV through emphasis on:
   (1) Requirements reform
      (a) No "overdesigning"
      (b) Prioritized tradable performance/mission requirements (threshold requirements, etc.)
   (2) Contractor configuration control, design flexibility
   (3) Commercial insertion/dual use, which is made possible by:
      (a) Mil spec reform
      (b) Government-industry IPTs

Deputy Under Secretary of Defense for Acquisition and Technology, presented a briefing with the following definition: "CAIV means that we will intentionally hold cost constant and accept the schedule and performance that results—within limits, of course."[3] According to a definition posted on the OUSD (A&T) Web site in early 1999, CAIV is "DoD's acquisition methodology of making technical and schedule performance a function of available budgeted resources." Most definitions, including Longuemare's, recognize that even under CAIV, cost is not an absolute fixed variable.[4] As Longuemare's

---

chronology of AR measures in "Baselining Acquisition Reform," *Acquisition Review Quarterly*, Winter 2000.

[3] Quoted in Office of the Deputy Under Secretary of Defense Acquisition Reform, *Cost as an Independent Variable: Stand-Down Acquisition Reform Acceleration Day*, May 1996.

[4] According to the Department of Defense *Defense Acquisition Handbook* (June 30, 1998), CAIV is a strategy that entails setting aggressive yet realistic cost objectives when defining operational requirements and acquiring defense systems and managing achievement of these objectives. Cost objectives must balance mission needs with projected out-year resources, taking into account existing technology, maturation of new technologies and anticipated process improvements in both the DoD and industry. As system performance and cost objectives are decided (on the basis of cost-performance trade-offs), the requirements and acquisition processes will make cost more of a constraint, and less of a variable, while nonetheless obtaining the needed military capability of the system. Although much discussion of CAIV is centered on

discussion suggests in his 1995 briefing, CAIV is an attempt to emulate the commercial sector concept of "must-cost" price goals when developing new items. Must-cost price goals in the commercial world are generally high-priority objectives but are not absolutely inflexible.[5]

The central aspect of CAIV is that it raises cost considerations to a priority level at least equal to, and more often even higher than the traditional military program considerations of system performance and development schedule. Elevating cost to a much higher level of importance naturally encourages the insertion of less expensive commercial parts and technology into weapon systems, according to CMI advocates. CAIV requires that both government and industry conduct rigorous and formal cost/benefit trade-off analyses from concept formulation through development and production. One purpose of such cost/performance trade studies is to identify the "knee of the curve" after which each marginal increase in capability or performance becomes increasingly expensive.

In March 1996, CAIV became official DoD acquisition policy and was to be applied to all major new acquisition programs. In June 1996, DoD Flagship Programs Workshops began examining specific pilot programs for testing out the implementation of CAIV.[6] In structuring most pilot programs, DoD officials attempted to develop strong incentives so that CAIV could be successfully implemented by both program officials and contractors. The key incentives are (1) high-

---

new systems, there is always opportunity for cost reduction. CAIV principles are applicable throughout a system's life cycle.

[5]The term "must cost" is now commonly used in the commercial aerospace sector and other commercial sectors to imply the high priority placed on achieving aggressive price goals. Subcontractors subjected to must-cost goals know that they risk losing their contract if they do not meet the aggressive price targets established by the prime contractor. However, it is always possible that a must-cost goal is too aggressive and cannot be met by any subcontractor under any reasonable conditions. In such a case, the must-cost goal would obviously be adjusted upward (or performance requirements reduced). Thus, must-cost goals might best be considered a term of art because of the apparent contradiction between "must-cost" and "goal." See Lorell et al., *Cheaper, Faster, Better?*, Chapter 6, "Lessons from the Commercial Aerospace Sector."

[6]Programs eventually chosen included EELV, Air Intercept Missile (AIM)-9X, Army Tactical Missile System–Brilliant Anti-Armor Munition (TACMS-BSAT), Preplanned Product Improvement Program (P3I), Multifunctional Information Display System (MIDS), JASSM, Crusader, JSF, and SBIRS.

priority unit price thresholds and targets that contractors must meet; (2) average unit procurement price requirements (AUPPRs) written into the Operational Requirements Document (ORD) and contractor procurement price commitment curves (PPCCs) with positive and negative performance incentives as a key aspect of the down-select process for engineering and manufacturing development (EMD) as well as the production contract; and (3) maintaining contractor competition during the R&D phase for as long as possible. More is said about these concepts and approaches below.

The factors that in theory either enable CAIV or make it possible are (1) requirements reform, (2) contractor configuration control and design flexibility, and (3) commercial insertion. Requirements reform requires:

- Close scrutiny of system requirements to separate "must-have" capabilities from those that are only "nice to have";

- Formulation of system requirements in terms of mission performance rather than detailed technical system specifications; and

- Thorough analysis of cost/performance trade-offs.

First and foremost, the government buyer—the services and the DoD—must have a clear and precise understanding of what the mission for the system is and what outcome is needed from this system on the battlefield. The buyer must then carefully prioritize the mission performance needs and broad capability requirements that the system should possess in order to accomplish the mission.[7] Prioritization is critical so that intelligent trade-offs can be made between cost and capability. A key objective of this approach is to avoid "overdesigning" weapon systems with higher performance and more extensive capabilities that are not truly necessary to perform the mission. Furthermore, AR advocates argue that overdesigning drives up costs by necessitating the use of unique military-only parts and technologies that cost far more than roughly equivalent commercial parts and technologies with perhaps slightly lower performance capabilities. Thus, requirements reform is a key element of

---

[7]These may include factors such as reliability, sortie rate, survivability, and robustness along with more traditional measures of performance such as speed, range, and payload.

CAIV and is necessary for the full exploitation of commercial technology, according to AR advocates.

Second, AR advocates argue that the service buyer should not dictate specific or detailed technical and design solutions to contractors. Instead, contractors should be provided only with those general system and performance requirements that are necessary to accomplish the military mission. As in the commercial world, defense contractors should be given much more opportunity to develop new and innovative design and technical solutions at lower cost in order to meet mission requirements.

For the CAIV process to achieve its full potential, according to AR advocates, two other conditions are necessary: contractor configuration control—at least below the overall system level—and commercial insertion. Configuration control combined with commercial insertion permits the contractor to seek out and experiment with any technologies and parts available in the marketplace, whether commercial or military, in order to meet government buyers' mission requirements at the lowest possible cost.

Finally, contractors and the government must conduct extensive cost/performance trade studies to determine at what point equal marginal improvements in performance become increasingly expensive. This analysis is necessary so that the user understands the cost of increasing performance in any given area and recognizes at what point the phenomenon of diminishing marginal returns comes into play. In this way, the user community can make more informed judgments regarding the prioritization of performance requirements.

Commercial insertion has been made possible by mil spec reform and by government-industry integrated product teams (IPTs). As was pointed out earlier, mil spec reform has been a key component of the DoD's AR policy since 1994. AR advocates argue that the wholesale application of mil specs to military programs inhibits the incorporation of less expensive and often more advanced commercial technologies and processes into military products while inhibiting the participation in military acquisition programs of commercial firms that use only commercial specifications and standards.

As mentioned earlier, former Secretary of Defense Perry issued a memorandum in mid-1994 to remedy this perceived problem. This memorandum, entitled *Specifications and Standards—A New Way of Doing Business*, turned existing DoD policy on its head: Instead of requiring mil specs, as had been the case in past policy, it called for the use of commercial and performance standards wherever possible and required that defense programs provide special justifications if mil specs were used. The services then individually reviewed mil specs in order to eliminate those which were unnecessary, substituted commercial standards when possible, or, where appropriate, updated existing mil specs.

Government-industry IPTs provide the trust and the constant communication necessary to permit industry to test and experiment with commercial parts, technologies, and designs that will meet government requirements, according to AR advocates.

These are the basic principles applied in the most comprehensive commercial-like AR pilot programs. We now turn to a detailed discussion to see how these principles are implemented to reduce costs.

It is extremely important to note that in the following sections on case studies of commercial-like acquisition programs, the claimed savings were estimated by comparing estimated projected costs before the imposition of AR measures with estimated projections after the imposition of such measures. Few are based on hard data. That is, few of the estimates contain "actuals," or actual cost data based on real work undertaken during product development and production. Most of the estimates were made before the beginning of system development or in the early phases of development. Even in cases where actuals were used in order to show claimed AR savings, the comparison of the actuals was made to an earlier estimate that is only a forecast and that itself is not based on actuals. Once a program has been restructured in the planning phase and launched as an AR effort, it is still possible to compare estimates or actuals only to a preprogram projection that was not based on actuals. In that sense, it is impossible to know with certainty what savings, if any, were provided by AR, unless the program is run twice—once with AR and once without it. In most of the cases reported below, the claimed savings estimates are based on comparing two projections, neither of which is founded on actuals. Therefore, these estimates must be viewed with extreme caution.

Chapter Four
THREE U.S. AIR FORCE ACQUISITION REFORM PILOT MUNITIONS PROGRAMS

MUNITIONS PROGRAM OVERVIEWS

We have chosen to begin our review of commercial-like pilot military procurement programs with three innovative acquisition efforts for the development and production of three different new "smart" munitions: JDAM, WCMD, and JASSM. The attractiveness of these programs is that they have been under way for some time and have thus provided some actual data. However, they also have many disadvantages and shortcomings as proof of savings that can be easily transferred to other types of programs.

To provide a clearer understanding of how the government sought to achieve major cost savings by structuring acquisition programs in a more commercial-like manner, we have provided detailed case studies of these three programs. The other seven pilot programs are examined in less detail, but most exhibit similar characteristics.[1]

JDAM is an important early trial program for testing out key aspects of the Clinton administration's defense AR measures. Indeed, in 1996, Lieutenant General George Muellner, then Principal Deputy Assistant Secretary of the Air Force (Acquisition), characterized

---

[1] Most of the information on these three programs included in this report was acquired from open published sources, from program documents, and from interviews conducted by the author with the Program Offices (all located at Eglin Air Force Base, Florida) and with contractors.

JDAM as "the linchpin" of "the broader Department of Defense's acquisition streamlining activities."[2] JDAM is an Acquisition Category (ACAT) 1D program, the most important DoD acquisition category.[3] JDAM is a joint Air Force/Navy program with OSD and Marine participation. The U.S. Air Force is the lead service.

JDAM originated as a traditional military acquisition program. Nonetheless, from the very beginning, the Air Force imposed a high-priority average unit price target of $40,000. In 1994, the DoD designated JDAM as an official DAPP under the 1994 Federal Acquisition Streamlining Act (FASA), which mandated a wide variety of AR measures.[4] Dr. Paul Kaminski, sworn in as Under Secretary of Defense for Acquisition and Technology in October 1994, strongly supported JDAM as a major test case for AR.

The JDAM program aims at developing sophisticated—but affordable—"strap-on" guidance kits that can be attached to standard Mk-83 and BLU-110 1000-lb. "dumb" bombs and to Mk-84 and BLU-109 2000-lb. "dumb" bombs. Through the use of an inertial navigation system augmented by updates provided by the Global Positioning System (GPS), which guide active control surfaces, JDAM kits permit highly accurate delivery of bombs from a variety of aircraft platforms under a wide range of adverse weather and environmental condi-

---

[2]Keynote Address, Orlando Air Force Association Symposium, February 16, 1996.

[3]ACAT 1D programs are MDAPs. According to the Defense Systems Management College, "An MDAP is defined as a program estimated by the Under Secretary of Defense (Acquisition and Technology) (USD [A&T]) to require eventual expenditure for research, development, test, and evaluation of more than $355 million (in fiscal year 1996 [FY96] constant dollars) or procurement of more than $2.135 billion (FY96 constant dollars), or those designated by the USD [A&T] to be ACAT I." ACAT 1D programs are those in which the Milestone Decision Authority (MDA) resides at the highest level possible: USD (A&T).

[4]The DoD Authorization Act for FY94 designated five programs as statutory DAPPs: JDAM, FSCATT, JPATS, Commercial Derivative Engine (CDE), and the Non-Developmental Airlift Aircraft (later dropped). FASA provided regulatory relief for these programs and gave authorization to treat them as commercial procurements. Later, the C-130J and the Defense Personnel Support Center (DPSC) were added as "regulatory" DAPPs. See Department of Defense, Pilot Program Consulting Group, *Celebrating Success: Forging the Future*, 1997, and Office of the Under Secretary of Defense, Acquisition and Technology, Acquisition Reform Benchmarking Group, *1997 Final Report*, June 30, 1997.

Three U.S. Air Force Acquisition Reform Pilot Munitions Programs   41

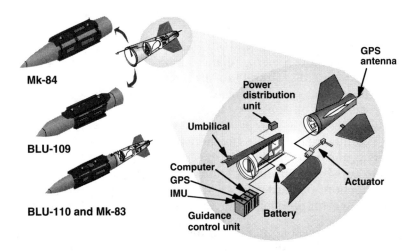

Figure 4.1—JDAM Baseline Weapons

tions.⁵ JDAM has a range of roughly 15 nautical miles when dropped from high altitudes. The JDAM configuration and baseline weapons are shown in Figure 4.1.

The Air Force WCMD program has some similarities to the JDAM effort. In response to FASA and the DoD's AR efforts, the Air Force designated WCMD a "lead program" to test out AR within the Air Force. WCMD is the only Air Force AR lead program for a totally military-unique combat weapon system developed from scratch.⁶ Compared to JDAM, WCMD is a somewhat simpler tail guidance retrofit kit employing an inertial navigation unit and active control

---

[5]U.S. and allied forces used a wide variety of existing "smart" munitions during Desert Storm combat operations in Kuwait and Iraq, often with great effect. However, many of those smart munitions guidance kits use electro-optical, laser, or infrared sensors whose performance can be degraded in poor weather conditions, when the battlefield is obscured by smoke and dust, or by other factors. The requirement for JDAM and WCMD arose from the need to develop munitions guidance kits for unguided munitions that could operate well in all weather conditions and in other situations where visibility is poor.

[6]WCMD is one of four Air Force lead programs selected to implement acquisition streamlining initiatives. The other three are EELV, Ground Theater Air Communications System (GTACS), and SBIRS.

surfaces intended for use on three "dumb" air-dropped munitions dispensers: the CBU-87/B Combined Effects Munition (CEM), the CBU-89/B Gator, and the CBU-97/B Sensor Fuzed Weapon (SFW). WCMD kits are intended to enhance aircraft survivability by permitting a GPS-capable aircraft to drop munitions dispensers from medium altitudes with accuracies equal to or better than those currently achieved through dangerous low-level attack profiles. WCMD's inertial measurement unit (IMU), which can be updated with GPS-quality data from the launch aircraft, corrects for launch transients and wind deflections, thus providing medium-altitude all-weather capability. Its active control surfaces and wind estimation and correction software help WCMD achieve a target accuracy of 85 feet circular error probable (CEP) from altitudes up to 45,000 feet.

JASSM is the largest and most sophisticated of the three programs. Like JDAM, JASSM is a joint Air Force/Navy project with the Air Force in the lead role. However, JASSM is a much more complex autonomous standoff munition. It is a long-range powered cruise missile with stealthy characteristics. Like JDAM, the missile is equipped with an inertial navigation system and a GPS receiver for navigation. In addition, JASSM will add a sophisticated autonomous terminal guidance and an automatic target recognition system for true standoff fire-and-forget capability. JASSM will have a range in the hundreds of miles depending on the launch platform and altitude. With overall performance objectives similar to the ill-fated Tri-Service Standoff Attack Munition (TSSAM), JASSM is a technologically challenging program, particularly in the areas of overall system integration, autonomous guidance, and automatic target recognition.[7]

The DoD approved the development of JASSM in September 1995, designating it a "flagship pilot program" for AR.[8] Former Assistant Secretary of the Air Force for Acquisition Arthur L. Money has characterized the JASSM program as employing "an aggressive acquisition approach using virtually every acquisition reform initiative

---

[7]See U.S. General Accounting Office, *Precision-Guided Munitions: Acquisition Plans for the Joint Air-to-Surface Standoff Missile*, GAO/NSIAD-96-144, June 1996.

[8]More accurately, JASSM is a "flagship pilot program for CAIV."

known to date."⁹ As an ACAT 1D program, JASSM, like JDAM, is also in the highest DoD acquisition category.

CAIV, Requirements Reform, and Must-Cost Targets on the JDAM, WCMD, and JASSM Programs

JDAM, WCMD, and JASSM are pioneering attempts at promoting lower cost through requirements reform. All sought to reduce costs by using carefully crafted mission requirements that avoid "overdesigning" and unnecessary capabilities; the presentation of system requirements to contractors in terms of mission performance rather than detailed design and technical specifications; minimal use of mil specs; contractor configuration control during R&D; and cost/performance trade studies.

In the official ORDs given to contractors, all three programs replaced detailed technical specifications and "how-to" design-and-build directives with broad mission performance objectives. These were for the most part prioritized into "key performance parameters" (KPPs), "critical performance requirements," and "threshold requirements." The purpose of this categorization was to focus contractor efforts on the most important program requirements and to facilitate and encourage trade-offs for cost and other reasons.¹⁰

Originally, program planners intended to require no mil specs whatsoever so that contractors could exploit off-the-shelf commercial technologies and parts to reduce costs. However, a few mil specs were eventually adopted to ensure compatibility with host aircraft as well as for safety considerations. For example, the weapon stores and software interface with the host aircraft required the use of MIL-STD-1760, while communications between the JDAM onboard processor and the host aircraft necessitated use of the MIL-STD-1553 high-speed bus. JASSM and WCMD experienced similar additions of some mil specs.

---

⁹Quoted by Suzann Chapman in "JASSM Competitors Chosen," *Air Force Magazine*, August 1996.

¹⁰These concepts are explained in greater detail later in this section.

Nonetheless, these programs still show a dramatic reduction in mil spec requirements compared to traditional programs. In the case of JDAM, the baseline pre-DAPP RFP included 87 mil specs compared to only a few in the DAPP phase. Interestingly, JDAM also did not require any specific commercial specifications or standards, nor were any mil specs or commercial standards embedded in its SOW, because only a statement of objectives (SOO) was required from the contractors. WCMD eliminated all but two mil standards.

In the area of performance requirements, all three programs remained very close to the original intent of using only broad mission performance requirements instead of detailed technical specifications. In the case of JASSM, the Air Force and Navy user communities agreed that only three KPPs were nonnegotiable: range, missile effectiveness, and aircraft carrier compatibility. Measures of merit for missile effectiveness were carefully developed and clearly communicated to competing contractors.[11] In addition, the government side developed seven "critical performance requirements." Many of these performance requirements had minimum thresholds that had to be met, but the critical performance requirements could still be traded off against each other and against other factors to reduce costs as long as the minimum threshold performance was met.

A fundamental program goal that is enshrined in the ORD is the "must-not-exceed" price ceiling of $700,000 (FY95) for the average unit procurement price of JASSM and the target price objective of $400,000 (FY95). The JASSM requirement emerged in 1995 after the cancellation of the TSSAM program. Begun in 1986 by Northrop, TSSAM aimed at providing a stealthy, long-range cruise missile with autonomous terminal guidance and target recognition capabilities. After many years of development, however, the Pentagon canceled TSSAM because the program was plagued with reliability problems and because of high costs. At the time of program launch, government officials estimated a unit production cost for TSSAM of $728,000 in then-year (TY) dollars. By 1994, average unit production costs for approximately 2500 missiles were expected to exceed $2

---

[11]The government gave the contractors the official JASSM ORD and used a computer simulation to measure missile effectiveness. All contractors had access to the model and could use it to test their technical proposals. Furthermore, contractors could test and question the methodology, tools, and assumptions built into the model.

million (TY dollars). OSD concluded that the program had to be canceled because of excessive cost and that a price above $700,000 (FY95 dollars) for a TSSAM follow-on missile would prevent the procurement of adequate numbers of missiles. The cancellation of TSSAM and the continuing critical need for an affordable long-range low-observable standoff missile with performance capabilities similar to TSSAM are the key reasons OSD designated JASSM a flagship program for CAIV.[12]

Yet the JASSM contractors were not told *how* to achieve critical performance requirements such as "missile effectiveness" while achieving the $400,000 average procurement price target. For example, one defined characteristic of the missile effectiveness KPP was survivability. This could be achieved by missile speed, lowering radar cross section, or a variety of other means. Alternatively, survivability could be traded off against other defined characteristics of missile effectiveness, such as reliability or probability of damaging various types of targets, or against cost (assuming that minimum survivability performance thresholds had been met). It was up to the contractor and its engineers to use creative new approaches to try to optimize the trade-offs between a variety of factors, meet the target price goals, and convince the customer that the correct design trade decisions had been made.

As mentioned above, critical performance requirements were further prioritized by including "threshold" requirements that were defined as very high priority together with "objectives" that were lower priority. Objectives could be traded off against cost and against each other. An example of a threshold requirement was missile compatibility with the B-52 bomber, the U.S. Air Force F-16 fighter, and the Navy F/A-18E/F.[13] JASSM missile compatibility with a variety of other aircraft was labeled an objective. Contractors carefully assessed the cost benefits of not achieving certain stated objectives,

---

[12] An important initial requirement that drove up costs on TSSAM but was dropped on JASSM was triservice deployment capability. TSSAM had to be capable of launch from both Air Force and Navy aircraft as well as from Army ground launchers. This requirement raised numerous technical difficulties for TSSAM developers. JASSM dropped the Army ground launch requirement and retained only the Air Force and Navy air launch requirement.

[13] The F/A-18E/F was later dropped as a threshold aircraft.

such as missile compatibility with the F-117, and even some threshold requirements, and discussed their findings with the government. This helped both the government and the contractors clearly understand the cost of achieving each operational capability and decide if that capability was really worth the added cost. This iterative process between contractors and the government led to changes in emphasis and priorities in overall system requirements.

The government structured the JDAM and WCMD design phases in a similar manner, making sure that the contractors knew which performance requirements were considered essential and which were more flexible, and emphasizing low cost as a key objective. In the case of JDAM, for example, the original ORD contained seven KPPs. The first six were grouped together as critical performance requirements. Some were absolute requirements, and some included minimum threshold performance requirements with greater trade flexibility once the minimum threshold had been achieved. These six were:

- Target impact accuracy of 13 meters CEP with GPS;
- Accuracy unaffected by weather conditions;
- In-flight retargeting capability (before release);
- Warhead compatibility;[14]
- Carrier suitability; and
- Primary aircraft compatibility.[15]

The seventh KPP was a ceiling on average unit production price. In the early 1990s, initial generic program estimates of average unit procurement price (AUPP) for 40,000 units of a JDAM-type weapon

---

[14]JDAM guidance kits had to work with the Mk-84 general-purpose 2000-lb. bombs, the BLU-109 2000-lb. penetrating bomb, and the BLU-110/Mk-83 general-purpose 1000-lb. bombs.

[15]Four aircraft (F-22, B-52H, F/A-18C/D, and AV-8B) were listed as "threshold requirement" aircraft (1000-lb. bomb versions only for F-22 and AV-8B), which means this capability requirement had a very high priority. Compatibility with nine other aircraft (B-1, B-2, F-16C/D, F-15E, F-117, F/A-18E/F, F-14A/B/D, P-3, and S-3) was listed as an "objective." Compatibility with the objective aircraft was a requirement fully subject to trade-off analysis with cost and other factors.

developed under a traditional acquisition approach ranged as high as $68,000.[16] Senior Air Force officials concluded that because of budget limitations, JDAM could not be procured in adequate numbers at this price. As a result, the seventh KPP was not negotiable or tradable. Placed prominently in the ORD, it required that JDAM have an AUPP of $40,000 or less.[17]

The updated 1995 version of the ORD raised the bar on cost even more by changing the requirement of an "AUPP of $40,000 or less" to the status of a minimum threshold and by designating an AUPP of $30,000 as the desired target price.

These seven KPPs were not tradable except at performance levels superior to the minimum threshold levels established for some of the critical performance parameters. The original ORD had many other requirements, but they were all tradable against cost or other factors and usually had no minimum threshold performance level that had to be achieved.

The focus on cost through CAIV, the use of broad mission requirements, the emphasis on cost/benefit trade-offs, the lack of mil spec requirements, and the control of the contractor over configuration and technical solutions seem to have produced dramatic results. Contractors often took the initiative to exploit commercial technologies, insert commercial off-the-shelf (COTS) parts and components, develop creative technical solutions, and trade off performance against cost where appropriate to achieve significant cost reductions.

According to Boeing St. Louis, more than 200 cases of detailed trade-off studies that reduced JDAM costs have been formally documented in the program's affordability trade studies, although most of the specific cases are proprietary. One of these cases is discussed below as an example of the insertion of dual-use technology.

---

[16]In 1993 dollars. The first official AUPP program estimates for JDAM were significantly below this figure. The highest levels of the U.S. Air Force imposed a strict $40,000 average price target on the JDAM program from its very inception. A more detailed discussion of initial program production cost estimates is presented in the cost section of this chapter.

[17]For 40,000 units in FY91 dollars. Using the official OSD inflator for USAF missile procurement, this amounts to $42,239 in 1993 dollars.

Because of the emphasis on cost promoted by CAIV, the trade-off analysis of performance versus cost that CAIV encouraged, and the elimination of most mil specs, JDAM is able to make extensive use of COTS processors, boards, chips, and other commercial parts and components. Originally, program officials and contractors had planned to acquire major subsystems and components from commercial sources or production lines; Table 4.1 lists the planned sources for various key components for the designs of the two competing contractors during the final competition phase. In the case of the Boeing design, the IMU, the GPS receiver, the mission computer, and the control actuators made up 85 percent of the cost of the guidance kit. Although these subsystems are now acquired from military production lines, all contain commercial parts, are slightly modified versions of commercial items, are government off-the-shelf items (GOTS), or could be sold as commercial items.

For example, the Boeing JDAM mission computer, as shown in Table 4.1, was originally intended to come from a commercial source. Eventually, however, Boeing designed its own mission computer and selected Unisys (now Lockheed Martin Tactical Defense Systems) to manufacture that computer on a military production line. Boeing's dedicated military mission computer is programmed using the Ada language, which is uncommon in the commercial world. Nonetheless, the mission computer's architecture is similar to that of desktop computers. At its heart is a Motorola microprocessor similar to the one that, prior to the JDAM program, was used by Apple Computer as the basis for its Performa 470 series of personal computers. Boeing hopes to upgrade this chip with one similar to that used in the PowerPC or iMac.

Both JDAM and WCMD use the Honeywell HG1700 IMU. This highly miniaturized, dedicated military IMU was developed by Honeywell Military Avionics explicitly for applications such as smart munitions, unpiloted air vehicles (UAVs), and missiles. Similar IMUs are used in commercial applications such as railroad vehicle control and landslide detection because of their low-cost and high-performance

Table 4.1
Commercial/Military Mix of JDAM Contractor Production Lines[a,b]

	Boeing[c]	Lockheed Martin
Integration/assembly	COM	MIL
IMU	MIL	MIL
GPS	MIL	COM
Mission computer	COM	N/A
Circuit cards	COM	N/A
Connector	COM	N/A
Actuators	COM	MIL
Power supply/distribution	MIL	COM
Thermal	MIL	MIL
Container	COM	MIL/COM
Fin	COM	COM
Tail	MIL	MIL/COM
Hardback/nose	COM	MIL/COM

[a]As of late 1996, Boeing, the winning contractor, later switched the mission computer to a military production-line source. Data were gathered from the "JDAM Industrial Capability/Financial Viability Assessment" in U.S. Air Force, *Single Acquisition Management Plan for the Joint Direct Attack Munition (JDAM)*, Eglin Air Force Base, August 23, 1995 (with updates as of February 10, 1997).

[b]MIL = military; COM = commercial.

[c]Formerly McDonnell-Douglas, winner of the Phase II contract.

characteristics. Not only is the HG1700 essentially an off-the-shelf military item from the JDAM and WCMD perspective, but, as part of AR, Honeywell and Boeing implemented 11 design changes, or affordability initiatives, that reduced production costs by some 20 percent. These cost reductions were passed on in part to the DoD.[18]

One specific example of cost reduction through commercial parts insertion, which is documented in one of Boeing's affordability trade studies, is the case of the Honeywell HG1700 IMU. On the JDAM program, a Boeing/Honeywell IPT worked hard to reduce the cost of this item through the identification of cheaper commercial parts for insertion into the IMU as well as through other reform initiatives. For example, the original HG1700 IMU connectors were expensive mil spec parts. Eventually a way was found to use much cheaper

---

[18]Some of the initiatives included changes in make/buy decisions, parts changes, investment in cost-saving capital equipment, and using commercial inspection processes. One such initiative is discussed in greater detail below.

Honeywell commercial IMU connectors, which saved roughly $100 per JDAM IMU. This change alone has the potential to save millions of dollars in production costs over the planned JDAM production run.

Both prime and subcontractors conducted extensive testing of non-mil spec commercial and plastic-encapsulated parts and their applicability to the environmental conditions in which JDAM would operate. A temperature range of –55°C to +85°C was eventually accepted as the baseline standard for electronic parts. On the high end, this standard permits the use of catalog COTS industrial or automotive-grade parts. However, the low end surpasses the requirements for commercial parts and is indeed the same as the mil spec standard. Therefore, commercial catalog parts usually had to be tested and/or screened.

According to one Boeing JDAM official, the contractor's experience with testing commercial parts for insertion into JDAM subsystems was highly variable. Some suppliers conducted their own testing for Boeing at a relatively low cost. Other suppliers were willing to conduct tests at their own facilities but charged Boeing a substantial premium. A third category of suppliers agreed to sell testing devices or data to Boeing so that the prime contractor could conduct its own testing. Again, depending on the part or subcomponent, Boeing's cost of testing the commercial parts itself varied considerably. Finally, some suppliers agreed to sell commercial parts but refused to conduct the additional testing required and would not provide the data or devices necessary for the prime to conduct the tests.

Boeing officials claim that the extensive trade studies and commercial parts testing conducted during the initial phase of the program to identify appropriate commercial parts for insertion into JDAM proved to be an expensive and time-consuming effort. Nonetheless, the extra effort necessary to qualify commercial parts seems to have paid off in much lower production costs. According to one account, the use of plastic-encapsulated parts saved $535 per unit.[19] This is roughly 3 percent or less of the AUPP of the JDAM in 1998.

---

[19] Assistant Secretary of the Air Force, Acquisition, *JDAM—The Value of Acquisition Streamlining*, no date.

Similar incentives instilled by the CAIV approach, in combination with the virtual elimination of the need to use mil spec parts and processes, yielded similar results on the JASSM program: extremely creative and innovative approaches to exploiting existing commercial and military technologies and parts to lower costs while still producing acceptable performance capabilities in a military environment. Two interesting examples on JASSM are the process technologies chosen to manufacture the fuselage as well as the wings and vertical stabilizer. The winning contractor (Lockheed Martin) wanted to make all these structural elements primarily out of nonmetallic composite materials in order to lower weight and enhance stealthiness. Experience suggested, however, that finished load-bearing military structural parts manufactured from traditional aerospace composite materials and processes generally averaged from $600 to $1000 per pound. Using these processes and materials could rapidly escalate the cost of JASSM past the target and even beyond the ceiling prices.

Lockheed Martin and its subcontractors began looking around for solutions in the commercial world. Eventually engineers began examining vacuum-assisted resin transfer molding (VARTM), a process used in making fiberglass hulls for pleasure boats. This process produces finished fiber composite parts that cost about $5 per pound. Although the resulting parts are not appropriate for aerospace applications, engineers experimented with variations on this process using different materials systems. Eventually an approach was discovered that, while more expensive than the VARTM process for boat hulls, turned out to cost only a fraction of traditional aerospace approaches that require high temperatures and pressures for curing and thus need to be processed in expensive autoclaves. The modified VARTM approach was used for the body of the JASSM. In addition, engineers developed a lower-cost automated braiding platform to lay down the fiber matrix for the body that was conceptually based on commercial machines used to braid socks, shoelaces, and freeway pillar reinforcement rings. The result was an estimated fivefold or greater reduction in cost.[20]

---

[20] Specifically, $120 per pound for the modified VARTM process compared to typical aerospace composite fabrication and assembly costs of $600 to $1000 per pound, according to Lockheed Martin Skunk Works personnel.

A similar approach was tried with the wings and vertical fin of JASSM. Here, Lockheed Martin adopted a variation of the process used by commercial firms to build surfboards and windmill blades for wind-driven electrical energy generators. This process uses an outer-composite shell and an inner foam core to form a durable, lightweight structure. Although the process had to be modified considerably, the contractor estimates that it reduced costs by a factor of ten compared to traditional aerospace composite structure costs.[21]

In the case of the JASSM engine, Lockheed Martin used a combination of approaches. First, in order to save development costs on a new engine, designers selected an existing GOTS engine that had been used to power the Harpoon antiship missile for two decades. Second, the prime contractor helped the engine vendor lower the cost of the engine by one-third by replacing outdated mil spec parts and technology on the engine with modern but much cheaper commercial parts and technology. For example, the old mil spec analog engine controller was replaced by a modern digital controller. The latter technology was based on an off-the-shelf automotive processor.

Many other automotive and industrial-grade non-mil spec parts were used. The prime contractor usually asked the subcontractors and vendors to qualify the commercial parts if extra testing was needed.[22]

In all these areas, the JASSM primes used another common mechanism from the commercial world to keep costs under control: must-cost price targets. Aggressive cost targets for each major subsystem and component, such as the guidance and control units, were provided to vendors. This in turn encouraged vendors to insert COTS parts and technology to keep costs down.

In cases where no existing commercial product existed to meet the need, JASSM engineers sought out existing military technologies and parts to avoid the expense of having to develop entirely new items.

---

[21] The savings result primarily from a large reduction in fabrication and assembly labor and inspection costs.

[22] WCMD uses a high percentage of non-mil-spec industrial or automotive-grade commercial parts.

In order to achieve its performance requirements for autonomous terminal target acquisition and guidance, JASSM needed to use advanced sensors with target recognition capability. No appropriate commercial technologies existed to meet these needs.[23] According to some published sources, however, Lockheed Martin and its subcontractors were able to develop a derivative of the Imaging Infrared (IIR) seeker developed for the Hellfire or Javelin anti-tank missile that is appropriate for JASSM. It is claimed that this seeker fills the basic requirement at a cost of only $50,000.[24]

In a similar manner, Boeing, the losing JASSM contractor, worked closely with its subcontractors to reduce costs by adopting existing GOTS military hardware where commercial technology did not exist. In the case of the terminal guidance system, Boeing adopted a derivative of the infrared seeker already used in the AGM-130-powered standoff weapon. Instead of developing a new subsystem, Boeing incorporated the guidance system for its JASSM design that it was already using for its JDAM kit and also used an antijamming GPS receiver already developed by the Air Force. Finally, Boeing's design made use of autonomous target recognition (ATR) software that had already been developed for its Standoff Land Attack Missile Expanded Response (SLAM-ER) missile under development for the Navy.[25]

In summary, the focus on CAIV required a conscious effort to avoid overdesigning and requirements "creep," the use of mission performance requirements, a heavy emphasis on must-cost pricing targets, contractor configuration control, mil spec reform, and cost-performance trade analysis. This use of CAIV and a must-cost commercial-like approach in turn encouraged contractors on these three munitions pilot programs to seek out commercial technologies and parts

---

[23]Early in the program, GAO identified the automatic target recognition requirement and autonomous guidance system on JASSM as areas of high technological risk that could cause schedule slippage and cost growth. More is said on this below. See U.S. General Accounting Office, *Precision-Guided Munitions: Acquisition Plans for the Joint Air-to-Surface Standoff Missile*.

[24]This seeker does not, however, provide true all-weather capability. It is limited to a 1500-foot ceiling and three-mile visibility. See "USAF to Begin Planning JASSM Upgrades," *Aerospace Daily*, April 29, 1998.

[25]David A. Fulghum, "Boeing Unveils Stealth Standoff Missile Design," *Aviation Week & Space Technology*, March 9, 1998.

that could lower costs while maintaining adequate performance or, if no commercial part existed, to incorporate existing GOTS military parts and subsystems. Contractors were thus able to offer the government a richly varied menu of cost/benefit trade-offs and alternative design solutions because the government provided no detailed system specification and did not demand the use of military specifications and standards. Indeed, the use of commercial parts, components, and processes was encouraged if they lowered costs and provided acceptable performance. The contractors were given almost total control over configuration, design, and technical solutions. If a commercial part slightly reduced environmental robustness, a contractor could still argue that the cost savings far outweighed the loss in capability. The result was that the system design, its expected capabilities, the cost estimates, the technical solutions chosen, and the suggested parts and components all came from and were "owned" by the contractors, not the government. Much lower costs than might be expected appear to have resulted from this approach. However, some potential doubts and problems remain, as discussed below.

Competition and Cooperation

The third key aspect of a more commercial approach to military procurement that promotes greater CMI is more commercial-like contractor selection. The three munitions programs under examination here focused on the following areas:

- Extended contractor competition during R&D;
- Government-industry cooperation: maximum sharing of information;
- Rolling down-select; and
- Past performance criteria in down-select.

AR advocates clearly perceived continuous and intense competition as the driving force in the commercial marketplace that pushed firms to lower prices, increase quality, and improve performance. If buyers are dissatisfied with the price or quality of one company's products, they can always turn to another. In traditional military procurement programs, competition lasted only during the initial concept development stage or a prototype demonstration/validation stage. Reformers hoped to maintain competition longer throughout the EMD stage.

Government reformers observed that highly successful commercial firms have often developed relatively open and trusting relationships both with key customers and with suppliers. Information is shared; problems are worked out together. This contrasts sharply with the traditional adversarial relationship between government and contractors. AR advocates hope that by bringing government and industry personnel together in Integrated Product and Process Teams (IPPTs) and other cooperative arrangements during the phase in which contractors are competing, program outcomes will be improved.

Reformers believe that a key aspect of this new cooperative relationship would be the "rolling down-select." This concept implies constant interaction between contractors and the government to help contractors identify the weaknesses in their proposals and develop the best possible offer for the government prior to down-selection.

Finally, government reformers argue that contractors are motivated to perform well in the commercial world because they believe they will be rewarded with new contracts for having performed well on past contracts and, conversely, will be denied future contracts for having done poorly in the past. Traditionally, government contracts have been awarded to firms whose proposals contain the lowest-cost estimates or that promise the highest capabilities, with less emphasis placed on an individual firm's past record in delivering on promises. The Past Performance Value (PPV) concept was developed to apply this commercial standard to the selection of military contractors.

With respect to the three munitions programs, government officials had hoped to fund at least two competing contractors through the entire EMD phase for at least one of the munitions programs.[26] It rapidly became clear, however, that this was not feasible from a cost standpoint. Instead, government officials adopted the following approach. First, considerable effort was made to attract as many competitors as possible—particularly nontraditional commercial con-

---

[26]The R&D phases of all three munitions programs are essentially funded by traditional military cost plus fixed fee (CPFF) or cost plus incentive fee (CPIF) contracts.

tractors—into the initial conceptual design stage. These contractors then took part in an initial low-cost paper competition. The actual EMD phase was then divided into two parts. The first phase was intended to focus on lowering development risk, reducing unit costs, and reducing manufacturing risks. The government funded two competing contractors during this phase. At the end of the first phase, the government selected one of the competing contractors to complete development. A major factor in the selection of the winner was the contractor's ability to achieve a low production price.

Thus, in the case of WCMD, at least eight contractors submitted serious proposals in the original design phase, including some companies that might not normally have entered a military system design competition of this sort. Five contractors competed in the initial design competition for JDAM before it was designated as a DAPP. JASSM received seven serious design proposals at the beginning of the program. Particularly in the case of WCMD and JASSM, the initial buildup period to a final RFP was characterized by intense and cooperative interaction between the government program offices and the competing contractors with respect to requirements, design concepts and approaches, and the like.

In April 1994, before JDAM became a DAPP, program officials followed fairly conventional procedures to select Lockheed Martin and McDonnell Douglas (now Boeing) to continue competing during the 18-month Phase I EMD contract. Lockheed Martin and Alliant Techsystems won the WCMD first-phase contract in January 1995, while Lockheed Martin and Boeing became the JASSM finalists in June 1996 for the program definition and risk reduction (PDRR) phase. WCMD and JASSM helped pioneer the concept of PPV during these competitions.

In the case of JASSM, past performance was assigned a weight equal to all other factors in the contractors' proposals for the down-select to the PDRR phase. JASSM officials developed five broad categories of contractor past performance: manufacturing and cost/schedule performance were weighted equally at 25 percent each of past performance; product, aircraft integration, and software performance made up the rest of the past performance categories. Similar or related products developed by the contractor in the past were examined. Thus, only past performance and capabilities of direct relevance to JASSM, such as aircraft integration and software development, were assessed. The outcome of this assessment was given equal weight to the content of the actual current proposal. In the JASSM system proposals, assessments of development and production costs were given equal weight to achievement of KPPs and other requirements.[27] The WCMD program office also assigned past performance a weight equal to the technical and cost elements of the contractor proposals as well as the risk assessment.

Perhaps most important, as part of the concept of a rolling downselect, the contractors were informed of significant weaknesses and deficiencies in their proposals and their past performance evaluations. Contractors had full access to the criteria, standards, and methodology used in the evaluations. Contractors had numerous opportunities to respond and discuss government criticisms of both the technical proposals and past performance.

The JDAM program office adopted the rolling down-select concept during its EMD Phase I. The two contractors had asked for and had received significantly different levels of funding because they had taken different technical approaches. The contractors were thus measured against their own SOWs rather than directly against each other. During the first year and a half, at six-month intervals, government officials provided the two competing contractors with detailed "report cards" on their proposals showing areas of strength and weakness. Again, the contractors understood the measures of merit and had a full opportunity to respond and even criticize the standards used if appropriate.

WCMD added an additional element to the down-select to its pilot production phase by conducting a live fly-off (or bomb-off?) using the two competing contractors' tail kits. The same F-16 carried one contractor's system on one wing and the competing contractor's system on the other so that the same conditions would apply to both. This directly competitive fly-off helped lead to the selection of

---

[27]JASSM had four cost categories that in total were weighted at 25 percent of total evaluation criteria: AUPP of Lots 1–5, mission cost effectiveness, total contract price, and AUPP of Lots 6–10. KPPs were assigned weighting equal to only 12.5 percent of the total. See Lieutenant Colonel Chris King, Air Force Program Executive Officer for Weapons Program, "Joint Air-to-Surface Standoff Missile (JASSM): Acquisition Reform in Action," unpublished briefing.

Lockheed Martin in January 1997 to conduct the next phase of the project.

Thus, as in the commercial world, competition remained the central tool used in these programs to try to ensure low price and high quality in programs where traditional regulatory safeguards had been removed. Indeed, these programs took special measures to level the playing field and intensify the competition, and the measures taken were based on greater openness and better communication.

One of the most fascinating innovations in this area was the concept of multiple integrated government-contractor teams during the competitive EMD stage. In the case of JDAM during Phase I EMD, the System Program Office (SPO) established three separate IPTs. Each contractor had its own exclusive government-industry IPT, while the SPO formed a third government-only core team. The two government-industry IPTs were completely walled off from each other and had no access to each other's data or documents (which were all color-coded and marked as source selection sensitive). Only the core SPO team had access to both government-contractor IPTs.

Most interestingly, the central goal of the government members on the two contractor IPTs was to do as much as possible to help their specific contractor win the competition. The government fielded teams of 10 to 12 military and civilian officials who "lived" at each contractor site, not for the purpose of auditing or checking up on the contractor but to help the contractor lower his costs and improve his approach as much as possible. Contractors were allowed to use government IPT members in any way they wanted. One contractor closely integrated the government members into its design and engineering groups, while the other used them more like consultants and advisers to clarify issues and problems. This concept was also meant to supplement the feedback provided by the periodic report cards issued by the core team during the rolling down-select process. In this way, the SPO hoped that both contractors would improve their proposals to such an extent during the EMD Phase I that it would be almost impossible to choose a winner.

The central objective was to provide the contractors with as much leeway and as much information as possible and then to let them compete against each other in a manner that mimicked what takes place in the commercial marketplace. In the end, this approach appears to have provided good results. The final source selection for EMD Phase II came down primarily to a question of procurement price commitments, and even then the decision was a very close call.

Procurement Price Commitments During R&D

To ensure the production of low-cost, high-quality, reliable, and maintainable systems, the government developed a strategy to structure the purchase and support of these three munitions in a manner that sought to achieve the benefits enjoyed by buyers in routine commercial transactions. The main elements of this strategy are:

- Fixed low-rate initial production (LRIP) procurement price commitments made during R&D;
- Competition to encourage low price commitments;
- A "carrot-and-stick" incentive system;
- System performance guarantees and warranties; and
- Full contractor responsibility for reliability and maintenance included in the system purchase price.

Government planners believed the bulk of the savings that would be generated by a more commercial-like acquisition approach on these three munitions pilot programs would accrue during the production phases. During the R&D phases, the government still paid up front for all costs and, indeed, incurred extra costs by supporting two contractors during the first phase of R&D. However, the central focus of the R&D programs was to develop effective systems with much-reduced production prices. For all three programs, the government initiated the R&D phase by providing the participating contractors with a production price goal and a production price ceiling beyond which the item would not be purchased. For all practical purposes, these goals and objectives were similar to airline must-cost requirements placed on commercial transport prime contractors. The production price commitments provided by the munitions contractors and the credibility of these estimates were central factors determining which contractor won the down-select at the end of the first phase of R&D. These prices tended to be far below the original government price goals. The problem for the government program managers then became how to incentivize the contractor to ensure that the production price commitments, as well as the performance and reliability guarantees, were met.

For all three munitions programs, government officials have chosen to use mechanisms meant to emulate certain aspects of the commercial marketplace. Two such mechanisms are PPCCs and warranties. The final contractor proposals for the second phase of R&D for these programs included fixed prices for low-rate production. In the case of JDAM, the competing contractors agreed to an AUPPR in FY93 dollars as part of the official system specification in the ORD that they themselves had written. The AUPPR had to include the cost of a full "bumper-to-bumper" warranty.[28] The AUPPR applies to Production Lots 1 and 2, which make up the LRIP phase. The system specification also included procurement price objectives for quantities in excess of 40,000 and 74,000 units, which in essence provided an estimate of the contractor's production learning curve. Thus, the contractors committed to a firm fixed price for the first LRIP lots at the beginning of full-scale development. Unlike most commercial customers, however, the government required that cost data be submitted to back up the AUPPR. However, the cost data requirements were greatly simplified compared to a traditional program and were limited to "only" 15 pages.

At the end of the first phase of R&D, the contractors in all three munitions programs also provided a good-faith estimate of the production prices for production lots following LRIP. In the case of JDAM, the contractors provided a PPCC for Lots 3–5 (a total of some 8700 units). The JDAM contractors also agreed to submit PPCCs for Lots 6–11 at the time of their Lot 4 final price proposals. The PPCCs are intended to be good-faith best estimates and are not contractually binding. In addition, the government required no cost data to support them. However, the contractors agreed to an extensive array of carrot-and-stick incentives to encourage attainment of the PPCC.

---

[28] More is said on warranties below.

If the contractor submits actual production price offers for post-LRIP production lots that are at or below the original PPCC, the contractor enjoys the following benefits:

- The contractor remains the sole production source for a negotiated number of lots. The government will not request changes in subcontractors.
- The contractor retains full configuration control as long as changes do not reduce performance or affect the safety of flight. Changes must be documented and reported to the government.
- If the contractor is able to reduce his production costs through the insertion of new technologies or other efficiencies, such savings are retained entirely by the contractor as additional profit.
- The contractor does not need to submit any type of cost or technical data to the government, assuming that performance, reliability, and delivery schedules are being met.
- The government will actively help the contractor reduce costs if requested but will not pay to implement changes.
- There is no in-plant government oversight or inspection of the contractor or subcontractors. All acceptance testing is done by the contractor in accordance with mutually agreed procedures.
- The contractor receives an incentive fee if the accuracy and reliability of production units exceed the specification.

The munitions contracts also contained "sticks" to protect the government from unsatisfactory performance by the contractor, particularly in price and system performance. These measures can be implemented by the government if the contractor submits a price bid for a production contract lot that exceeds the PPCC. First, however, there is a grace period during which the contractor can explain the reasons for exceeding the PPCC. If the government decides not to accept the explanation, the following measures can be taken:

- The contractor must submit fully compliant certified cost and pricing data in accordance with TINA and other regulations.
- The government may reestablish control over configuration.

- The contractor must prepare and provide a fully compliant mil spec data package free of charge within one year.

- The contractor must fully qualify, at his own expense and within 12 months, a new contractor as a second source for production. Full qualification is defined as delivery and acceptance of a production unit by the second-source contractor and ten successful flight tests. The government may impose fines of $20,000 for each working day up to a total of $5 million for failure to meet these requirements within schedule.[29]

- The government may impose in-plant oversight and testing.

- The incentive fee option is eliminated.

In principle, potentially the most undesirable "stick" from the contractor's perspective is the requirement to qualify a second source at his own expense. This stick is an attempt to simulate the incentives in the commercial world, where in most cases an unsatisfied buyer has to option of turning to a competing supplier of the same or a similar product. This option encourages the contractor to fulfill his promises to the buyer. In the case of unique military hardware, especially when the government does not control the data package, the existence of other suppliers of nearly identical items is highly unlikely. Therefore, the penalty to contractors for failing to meet the promises to the government buyer on the three munitions program is that they must create the new supplier at their own expense—a severe penalty indeed, and presumably a strong incentive to perform as promised.

At least some JDAM contractor representatives view the reality somewhat differently, however. According to one contractor representative, in actual practice the formal incentives against price gouging become relatively weak by Production Lots 6–11, which represent the vast bulk of all production. From the contractor's perspective, a hard fixed-price commitment clearly exists for LRIP Lots 1 and 2, and a somewhat softer commitment exists for Lots 3–5. In the contractor's view, however, it would be difficult for the government

---

[29]Some of the specific terms vary for each system or may have been amended. For example, the WCMD contract apparently permits the contractor 18 months to qualify a second source.

to enforce the requirement to qualify a second source if there are problems with the PPCC, particularly for Lots 6–11, primarily because of issues related to proprietary data. According to one contractor representative, there really is no credible element among the contractual sticks that prevents price gouging in Lots 6–12.

On the other hand, the contractor representatives insist that another strong incentive exists to hold to the PPCC and not price gouge: reputation and the trend toward using past performance in future contract awards. This incentive, they argue, works extremely well in the commercial aerospace world. Thus, they maintain, good faith and past performance are the keys to protecting the government from price gouging.

In addition to cost, government planners were also concerned about system performance, including reliability. The government decided to provide three types of incentives to ensure that the contractor achieved system performance goals. These included:

- A commercial-style "bumper-to-bumper" warranty that includes system performance, reliability, and support;
- Linking receipt of the PPCC incentive "carrots" to achievement of the performance specification; and
- Establishment of a formal dispute resolution process.

From the beginning of the program, both competing contractors accepted the concept of a commercial-style warranty requiring that contractors meet system specification requirements. The terms of the warranty are flowed down to the major suppliers and vendors by the prime contractor. In the commercial transport industry, prime contractors often provide specific average performance and reliability guarantees that entail cost penalties to the prime contractor if they are not achieved. Airlines sometimes try to negotiate the trade-off of some warranties and performance guarantees for lower system prices. In the case of JDAM, there is no explicit cost penalty for not meeting specification requirements. However, unless the contractor's kits meet the full specification requirement as determined by the government customer, the contractor does not enjoy the benefits of the PPCC "carrots."

More specifically, the JDAM warranty, which is similar to those for the other two munitions programs, requires that the contractor replace or repair any JDAM kit that does not meet the system specification requirement or that contains defects in materials or workmanship as determined by the government buyer. The warranty remains in force for 20 years as long as the kit remains in its shipping container and for five years outside the container. If the kit is properly repacked in its container, the 20-year warranty goes back into effect. The warranty also applies to 50 hours of carriage life on the pylon of a combat aircraft and includes a specific number of on-off operating cycles of the system during flight.

Many of the aspects of the warranty are similar to the standard commercial transport rules for aircraft on ground (AOG) owing to a broken part. The warranty requires that the contractor ship out repaired or nondefective kits within a specific time period (within one business day for the early low-rate production lots). The contractor must pay for the cost of shipping to any place in the world. The warranty is not unconditional, however; for example, it does not cover combat damage, uncontrollable events, misuse, or abuse by the government. On the other hand, in at least one of the munitions programs, neither the contractor nor the government expects detailed records to be kept on specific kits that can prove how long the kit has been out of its container or how many hours it has flown on a pylon. In other words, implementation of the warranty is predicated on an expectation of good-faith intent on both sides.

However, the munitions contracts also include provisions for a formal third-party dispute resolution process if the government and contractors disagree regarding the application of the warranty or other aspects of the contracts. This process entails the use of a dispute resolution board (DRB) made up of three members who do not represent either party. Each party chooses one member from a list of five candidates provided by the other party, and these two members choose the third member. Acceptance of a DRB finding is voluntary for both parties. However, all opinions and materials used in a DRB proceeding can be used in traditional dispute resolution procedures or in litigation.

Ultimately, however, the most important enforcement mechanism for the warranty is the same as that in the commercial world: reputation and past performance. If a contractor refuses to honor a warranty obligation that the government customer believes is clearly legitimate, this behavior will become part of the contractor's past performance record, which will be evaluated in competitions for future system development programs. This is the same incentive that encourages companies in the commercial world to honor their performance claims and warranty commitments. For this incentive to be effective, however, there must be more than one credible source or contractor for future competitions, and past performance criteria must be an important element in down-selections.

In summary, these three munitions programs have been structured in a radically different manner from traditional programs in order to mimic the market incentives of the commercial world, promote the insertion of commercial technology, and reap the claimed cost savings and efficiencies that are prevalent in the commercial marketplace. Although these three programs are still in their early stages, they appear to be achieving many of the hoped-for benefits of the more rapid development of lower-cost, more effective weapon systems.

Munitions Program Outcomes

Cost. The three munitions programs were all structured in a way that mimicked the emerging must-cost environment in the commercial transport sector. At least in the cases of JDAM and JASSM, the government customers established must-cost maximum price thresholds above which the system would not be purchased. Later, contractors were encouraged, through intense competition and the application of CAIV, to develop aggressive price targets that were considerably below the maximum price thresholds. Finally, the contractors committed to meeting LRIP production price objectives and accepted a series of carrot-and-stick contractual incentives to ensure that the price goals would be met.

JDAM. As noted above, early program estimates for a JDAM-type weapon kit were as high as $68,000 in 1993 dollars for an AUPP for 40,000 units. However, the first official government estimates were in the $40,000 range. Table 4.2 shows the first official program cost estimates from the Air Force and from the OSD Cost Analysis Improvement Group (CAIG), which were developed prior to AR and prior to the designation of JDAM as a DAPP.[30] As these figures show, both the Air Force and OSD agreed in 1993 that in a traditional pre-AR environment, the AUPP for JDAM would be in the range $43,800 to $46,300 in 1993 dollars.

After the DoD designated JDAM as a DAPP, a must-cost threshold of $40,000 (1993 dollars) or less per unit was established. This must-cost goal was easily achieved and, once AR was implemented, was greatly surpassed. Two additional pre- and post-DAPP estimates of JDAM program costs and AUPP are shown in Table 4.3.

Table 4.2

First Official Pre-DAPP JDAM Program Cost and AUPP Estimates, August 1993 (1993 $)

Cost Element	Air Force Estimate	OSD CAIG Estimate
EMD	338M	330M
Aircraft integration and USG provided assets	737M	852M
Procurement	3245M	3428M
Operations and Support	277M	277M
AUPP (based on 74,000-unit buy)	43.85K	46.32K

SOURCE: Office of the Director, Program Analysis and Evaluation, Cost Analysis Improvement Group, Memorandum for the Chairman, Conventional Systems Committee, *CAIG Report on the Life Cycle Costs of the Joint Direct Attack Munition (JDAM) Program*, August 25, 1993.

---

[30]The CAIG resides within the DoD's Office of the Director, Program Analysis and Evaluation. The CAIG has a statutory requirement to develop an estimate of the life cycle costs of an acquisition program separate from that of the sponsoring service and present it to the Secretary of Defense at each program's milestone.

Table 4.3
Pre- and Post-DAPP JDAM Program Costs and AUPP ($M)

Cost Element	PB FY95	PB FY97	CAIG I (FY95$)	CAIG II (FY95$)
R&D	549.7	462.9	346	380
Aircraft integration	TBD	TBD	893	478
Procurement	4874.9	2062.8[a]	3593	2012
Operations and Support	TBD	TBD	290	130
Total Cost	5558.8	2525.7	5122	3000
AUPP[b]	65.9	23.4	48.6	24.4

SOURCE: Based on data from the Office of the Deputy Under Secretary of Defense, Acquisition Reform, Pilot Program Consulting Group, *PPCG 1997 Compendium of Pilot Program Reports*, pp. 1–4. CAIG MS I includes 380 kits; MS II includes 630 kits.

[a] Assumes a total procurement of 87,496 units. All other numbers assume a total procurement of 74,000 units, except for the AUPP numbers, which assume 40,000 units. Current total production is expected to be 89,000 kits plus foreign sales.

[b] Thousands of dollars; 40,000 units. AUPP for PB FY95 and PB FY97 is stated in TY dollars.

The first two columns in Table 4.3 are the President's budget (PB) in TY dollars. The second two columns are estimates in constant FY95 dollars projected by the CAIG. The pre- and post-DAPP numbers from the PB and the CAIG projections for comparable categories differ because of different definitions and assumptions and because the numbers were generated at slightly different times. Nonetheless, both show a decline in AUPP for JDAM of at least 50 percent from the pre-AR numbers (columns one and three) to the post-AR numbers (columns two and four). Both sets of estimates show a 40 to 50 percent decline in total program costs for JDAM after AR. In addition, although not shown in exactly comparable terms, the AUPP numbers are considerably below the $40,000 must-cost threshold established at the beginning of the program. In contrast, JDAM R&D savings as shown in the DoD *1997 Compendium of Pilot Program Reports* (Table 4.3) appear small or nonexistent.

Excluding aircraft integration costs, the PB numbers show roughly a 15 percent savings. However, the CAIG projections show an increase in R&D of approximately 10 percent. The CAIG numbers indicate almost a 50 percent decline in aircraft integration costs, but this improvement arose primarily from a reduction in the number of "threshold" aircraft requiring integration. On the positive side, the CAIG projections estimate more than a 50 percent decrease in operations and support (O&S) costs.

Table 4.4 presents evidence from more recent published sources at the U.S. Air Force Air Staff using slightly different data. These data show a slightly smaller savings on development at just under 15 percent. On the other hand, they indicate an even larger decline in AUPP to less than one-half the cost in FY93 constant dollars even when shown in then-year prices unadjusted for inflation.

The most recent published and unpublished sources suggest that in 1998, the AUPP for JDAM in FY93 dollars stood at roughly $15,000 and that the then-year dollar AUPP in FY98 stood at approximately $18,000. However, the resolution of some technical problems that were detected in 1997 during development and testing may lead to a real increase of 4 to 5 percent in both development costs and AUPP. According to one published source, the added cost to the JDAM unit price in FY98 dollars is roughly $850.[31]

The bottom line, however, is that in constant FY93 dollars, the 1998 AUPP remains considerably less than one-half the procurement price estimated before the program became an AR pilot program (see Table 4.5). With a total buy now projected on the order of 89,000 units, this results in an inflation-adjusted procurement cost savings to the U.S. government of at least $2 billion.

Table 4.4

Pre- and Post-DAPP JDAM Development Cost and AUPP[a]

Cost Element	PB FY95	PB FY99
Development ($M)	549.7	469.3
AUPP[a] ($K)	42.2 (FY93)	<20

SOURCE: Office of the Assistant Secretary of the Air Force, Acquisition, *Acquisition Reform Success Story: Joint Direct Attack Munition (JDAM)*, December 1997.

[a]40,000 units.

---

[31]"Navy Wants Upgraded JDAM for No More Than $50,000," *Aerospace Daily*, August 26, 1998.

Table 4.5

Estimated EMD and Production Percentage AR Savings for JDAM

R&D	Production
~15[a]	>60[b]

[a]Comparison of December 1997 SAF/AQ R&D PB FY95 and PB FY99 cost estimates with rough adjustments for inflation and cost growth due to recent technical problems.

[b]Pre-DAPP AUPP estimate compared to current reported AUPP in constant dollars.

WCMD. AUPP savings for WCMD on a percentage basis roughly equal those of JDAM when initial pre-AR estimates are compared to post-AR estimates. At the beginning of the program, the AUPP for WCMD had been projected at $25,000 in 1994 constant dollars for 40,000 units. This price included the average field installation unit price, which covered contractor installation of the kit in the field. As of mid-1997, the 1994 constant-dollar AUPP stood at $8937—a full 64 percent below the original must-cost price.[32]

In late 1996, with R&D nearly complete, Air Force officials estimated a cost savings on EMD for WCMD of 35 percent due to AR. This estimate was based on comparing the initial government estimate of supporting two contractors at a cost of $65.6 million (TY dollars) to a projected total EMD cost of $42.9 million (TY dollars). Unfortunately, a year later (in late 1997), several technical problems were identified during testing that required correction, and developing the technical fixes led to a small increase in total EMD costs. Published sources claim, however, that the contractor agreed not to increase the AUPP.[33] Table 4.6 also shows a highly subjective breakout of percentage contribution by various AR measures for the WCMD program. The percentages assigned to various AR factors are basically informed guesses made by program officials, so they are not readily verifiable independently. Moreover, the specific categories are not defined, so it is not clear how these categories fit into our taxonomy.

---

[32]Office of the Assistant Secretary of the Air Force, Acquisition, *Acquisition Reform Success Story: Wind Corrected Munitions Dispenser (WCMD),* June 12, 1997.

[33]See "Wind Corrected Munitions Dispenser Price Holds Despite Fixes," *Aerospace Daily,* March 23, 1998.

We have arranged them into our two categories of regulatory and oversight compliance cost premium and commercial program structure.

JASSM. Finally, although still in an early stage of development and experiencing some test problems, JASSM also appears to be fulfilling the promise of a more commercial-like acquisition approach by greatly surpassing its original goals for low-cost pricing. JASSM began with a must-not-exceed ceiling average unit price goal of $700,000 in FY95 constant dollars, and a target price goal of $400,000 in FY95 constant dollars, for a production run of 2400. The $700,000 price ceiling goal was also confirmed by a CAIG estimate. Government analysts estimated total development costs in FY95 constant dollars at $675 million.

Table 4.6

Estimated EMD and Production Percentage AR Savings for WCMD by Category[a]

		R&D		Production	
		Percent of Total R&D Savings	Percent of R&D Dollars Saved	Percent of Total Production Savings	Percent of Production Dollars Saved
(1)	Regulatory burden (CDRL reduction)	10	3.5	5	3.2
(2)	Commercial program structure/CAIV	90	31.5	95	60.8
	CAIV	20	7	40	25.6
	CCC[b]	45	15.8	30	19.2
	No mil specs	5	1.8	20	12.8
	IPT	20	7	5	3.2
Total		100%	35%	100%	64%

SOURCE: *Acquisition Reform Cost Savings and Cost Avoidance: A Compilation of Cost Savings and Cost Avoidance Resulting from Implementing Acquisition Reform Initiatives,* AFMC draft report, Wright Patterson AFB, Dayton, OH, December 19, 1996.

[a]Compared to a traditionally structured program.

[b]CCC = contractor configuration control, a composite of the following categories used in the AFMC draft report: Configuration Control, Total System Performance Responsibility (TSPR), SOO vs. SOW.

In early April 1998, the Air Force down-selected to one contractor to complete the development of JASSM. The winning contractor, Lockheed Martin, committed to an AUPP for the first 195 missiles of $275,000 in FY95 constant dollars—more than 30 percent below the target price of $400,000 and more than 60 percent below the threshold ceiling price of $700,000. Boeing, the losing contractor, also came in with an offer under the target price with an AUPP of $398,000 for Lot 1.[34]

The development phase is also expected to cost approximately 30 percent less than original projections and far less than the amount spent on the failed TSSAM program, as shown in Table 4.7. The contracts awarded to the two contractors for the JASSM initial PDRR phase totaled $237.4 million,[35] and the full-scale development phase was expected to cost on the order of $200 million. An early restructuring of the development schedule, as discussed below, led to an estimated increase in EMD costs to some $240 million. More recent problems may cause EMD costs to increase further. The final EMD costs, however, are likely to remain well below the original 1995 projections of $675 million (FY95 dollars). Indeed, as late as March 1999, senior OSD officials testified to Congress that the estimated program costs for JASSM still showed an AR savings of 44 percent.[36]

In conclusion, claimed cost savings from AR on these three munitions programs range from 15 to 35 percent on R&D, and from 31 to 64 percent on AUPP, as shown in Table 4.8.

---

[34]See "JASSM Beats Cost Target by 40%," *Aerospace Daily*, April 30, 1998. Recent accounts report the price has risen to $317,000 in FY95 constant dollars owing to a decrease in the size of the initial buys. Yet this AUPP is still more than 20 percent below the original target price. See "JASSM Cruise Missile Crashes in First Flight Test," *Aerospace Daily*, April 21, 1999.

[35]"Competing JASSM Contractors Chosen," *Air Force News*, June 1996.

[36]Statement of Stan Z. Soloway, Deputy Under Secretary of Defense, Acquisition Reform, House Armed Services Committee, March 2, 1999.

Table 4.7
Estimated JASSM R&D and Production Percentage AR Savings

R&D (PDRR + EMD)	Production Price
29[a]	31 (61)[b]

[a] Cost of the PDRR phase for two contractors plus projected EMD costs for restructured program (1998) compared to original pre-AR R&D estimates of total R&D program costs.

[b] The first percentage is the reported Lockheed Martin AUPP proposal for initial production lots compared to the AUPP program target of $400,000. The percentage in parentheses is the reported AUPP compared to the initial threshold program ceiling price of $700,000.

Performance and Schedule. Probably the single greatest concern of the opponents of a more commercial-like acquisition approach is that the elimination of regulatory safeguards and the insertion of commercial technologies into weapon systems will result in inadequate performance or performance shortfalls. For the most part, the three munitions pilot programs under consideration here do not seem to indicate that these concerns are warranted, although certain technical difficulties have raised some red flags about the compatibility of commercial-like trade-offs between flight safety and cost reduction and the DoD's traditional desire to overdesign systems to ensure high margins of safety.

JDAM has experienced several high-visibility technical problems during its aircraft integration testing. Most of these have been solved without much difficulty. Early in the flight test program, for example, some problems were experienced with radio frequency components and with the GPS systems. Later testing showed that the 2000-lb. BLU-109 and the 1000-lb. Mk-83 versions of JDAM were unstable at

Table 4.8
Summary of Estimated Percentage Savings for U.S. Air Force Munitions AR Lead Programs

	R&D	Production (AUPP)	Estimate Quality
JDAM	15	>60	Analysis
WCMD	35	64	Analysis
JASSM	29	31	Analysis

high angles of attack—a problem that reduced the delivery envelopes for both weapons. Solving this problem required a redesign of the aerodynamic strakes attached to the sides of the bomb as well as the redesign and retesting of flight control system software.[37]

During the JDAM flight test program, engineers also found that unanticipated system vibration was causing problems in the transfer alignment of the IMU. The problems arose only with the Mk-84 2000-lb. variant of the JDAM kit and only when it was mounted on the inboard pylons of an F/A-18 Hornet operating at low altitudes and high speeds. This concurrent combination of kit type, aircraft type, mounting position, altitude, and speed is quite unlikely, especially given that the F/A-18's inboard pylon is typically used for fuel tanks and not for weapons. Boeing had not designed JDAM for such a scenario. Nevertheless, Boeing was able to fix the problem by modifying the IMU's vibration isolator ring and sculling algorithm.

When the JDAM test units were then subjected to this high-dynamic-load region for more extended periods of time, however, it was found that the commercially derived friction brake could not withstand the unexpectedly high aerodynamic forces. Given that the friction brake is used to hold the fins steady prior to launch, the result was fin and fin-shaft fatigue from excessive vibration and movement. Once again, this caused problems in the transfer alignment of the IMU and, worse, caused fins to move or even fin shafts to break prior to aircraft separation.

Boeing's initial attempts to solve the friction brake problem proved inadequate and were eventually discarded. Boeing engineers then adopted an entirely new approach based on a positive fin-locking mechanism that "nails down" the fin until launch by inserting a metal pin into a hole in the fin. The pin retracts into the tail kit within one second when the JDAM-equipped bomb is dropped. In addition, the fin shafts and other parts had to be strengthened.

The additional nonrecurring engineering costs and the need to use more expensive parts during production have resulted in a 4 to 5 percent increase in EMD costs and in AUPP, as mentioned above.

---

[37]Department of Defense, Office of the Director, Operational Test and Evaluation, "Joint Direct Attack Munition (JDAM)," *FY97 Annual Report*, February 1998.

This is not trivial with a buy of approximately 89,000 units; the additional procurement cost is on the order of $75 million or more. Nonetheless, the JDAM price is still well below the threshold and target prices established at the beginning of the program.

Were the JDAM technical problems caused by the use of commercial parts and technologies as part of CAIV? The direct answer appears to be no. Although the friction brake that proved inadequate was an inexpensive commercial derivative item, its inadequacy probably arose from Boeing's failure to calculate correctly the magnitude of the dynamic forces to which the JDAM Mk-84 tail fins would be subjected under certain special conditions. However, this problem occurred in part because Boeing placed heavier emphasis on cost reduction than on designing for a low-probability worst-on-worst-case scenario. It could thus be argued that the commercial-like approach taken by Boeing was incompatible with the DoD's desire to achieve high margins of safety.

Interestingly, recent press reports indicate that the "underwing environment" of the F/A-18E/F "puts a lot of stress on the weapons it carries."[38] In other words, some of the problems encountered by JDAM may be attributable to unanticipated issues related to the unique aerodynamics of the F-18.

In addition, both WCMD, developed by a different contractor, and the Joint Standoff Weapon (JSOW),[39] which is not an AR pilot program, experienced similar problems during development, although not just on the F-18. During testing in late 1997, WCMD showed fin vibration and flutter problems when carried on an F-16 at supersonic speeds. Lockheed Martin engineers concluded that they had to use the same type of fix as that which Boeing engineers developed for JDAM: a positive fin-locking mechanism. The Air Force also encoun-

---

[38]Quoted from Robert Wall, "Lingering Concerns Stalk F/A-18E/F," *Aviation Week & Space Technology*, February 28, 2000.

[39]Developed by Raytheon Texas Instruments, JSOW is a winged standoff, unpowered precision glide munition that comes in three variants, all of which deliver unitary or submunition warheads of approximately 1000 lbs. It has a range of 15–40 nautical miles depending on launch altitude. Like JDAM and WCMD, JSOW is guided by a GPS link and by an onboard IMU. Like JASSM, one planned JSOW variant (AGM-154C) will have an IIR terminal seeker. The program is a joint Navy/Air Force effort led by the Navy.

tered problems with the WCMD autopilot software during testing in late 1997. This problem was resolved fairly quickly.[40]

In most areas that are not affected by technical developmental problems, JDAM and WCMD seem to have already met or exceeded their critical performance and reliability requirements. Probably the single most important requirement for these two weapons is accuracy. JDAM started with a 13-meter CEP requirement. During developmental and operational testing by the Air Force in late 1996 and early 1997, JDAM achieved an average CEP of 10.3 meters. By late 1998, one source claimed that JDAM was achieving an average of 9.7 CEP with an actual average miss distance of 6.5 meters.[41] Because of the success of the initial developmental tests, the Air Force authorized LRIP in April 1997.

The true test for JDAM, however, came during the extended air campaign over Kosovo in early 1999. Between late March and early May 1999, six B-2s delivered in excess of 500 JDAMs against targets in Kosovo, amounting to 11 percent of the total bomb load dropped by U.S. forces during this period. Taking advantage of the GPS-Aided Targeting System (GATS) on B-2s, JDAMs reportedly scored an average CEP of 6 meters, compared to the original 13-meter requirement.[42]

WCMD started with a threshold-accuracy CEP requirement of 100 feet and a target CEP of 80 feet. WCMD has consistently achieved accuracies that greatly exceed the target CEP in developmental testing with launches at subsonic speeds. During testing in mid-1998, WCMD is reported to have achieved miss distances of 5 to 30 feet. It is for this reason that the Air Force approved LRIP in August 1998.[43] Once the fin-locking mechanism is installed in later production lots,

---

[40]"USAF Has Fix for One WCMD Problem," *Aerospace Daily*, February 20, 1998.

[41]"Boeing Presses 500-Pound JDAM Kit for U.S., International Buyers," *Aerospace Daily*, September 22, 1998.

[42]The most infamous example of JDAM's remarkable accuracy came when B-2-launched JDAMs precisely hit a building at the heart of a dense urban area in Belgrade. Unfortunately, the U.S. government had misidentified the building. Instead of an important Serb target, it turned out to be the Chinese Embassy. See Bill Sweetman, "Coming to a Theatre Near You," *Interavia Business and Technology*, July 1999.

[43]"USAF Approves WCMD for Low Rate Production," *Aerospace Daily*, August 4, 1998.

76   An Overview of Acquisition Reform Cost Savings Estimates

accuracy with launches at supersonic speeds is also expected to meet or exceed the initial requirement.

JDAM's technical problems have led to a restructuring of the developmental and operational test programs, the production program, and a delay of roughly a year in the procurement of the BLU-109 2000-lb. bomb variant. In April 1997, the Air Force authorized LRIP of 900 JDAM kits for the Mk-84 bomb. Confidence in the weapon was so high that in 1997, Boeing delivered 140 "early operational capability" JDAM kits to the operational B-2 wing at Whiteman Air Force Base. The Air Force had originally planned to enter into full-rate production in 1998 with both the BLU-109 and the Mk-84 2000-lb. bomb kit variants. In late 1997, however, the Air Force delayed full-rate production and substituted a second lot of low-rate production made up exclusively of Mk-84 variants. The purpose of this change was twofold: to permit additional flight testing to work out the flight instability problems encountered with the Mk-83 and BLU-109 JDAM kits, and to continue the development of the fin-locking mechanism necessary to qualify the Mk-84 for the F/A-18 inboard pylons. Air Force officials claim that this change will have little effect on the production program, since approximately the same number of kits in the same bomb-size category will be procured as originally planned in 1998.[44] Because of the Kosovo air war, JDAM production was increased by 50 percent in May 1999.[45]

The WCMD technical problems led to a similar restructuring of the operational test and production phases of the program. The Air Force had originally planned to authorize LRIP in February 1998. Following the discovery of the supersonic launch fin flutter and autopilot software problems in November 1997, the Air Force stopped operational testing and delayed LRIP until fixes could be found. The Air Force, however, was determined to maintain the schedule for ini-

---

[44]See "Problems Force Delay in JDAM Full-Rate Production," *Aerospace Daily*, December 15, 1997, and "USAF Will Buy Only Mk.84 JDAMs This Year," *Aerospace Daily*, December 17, 1997. The BLU-109 is designed to penetrate and destroy harder targets than the Mk-84, so some capability will be lost. Also, the first two LRIP production lots of Mk-84 JDAMs will not have the fin-locking mechanism fix, so they will not be usable on F/A-18s. See also Department of Defense, Office of the Director, Operational Test and Evaluation, "Joint Direct Attack Munition (JDAM)."

[45]See "Boeing to Deliver Additional JDAMs Two Months Early," *Aerospace Daily*, May 3, 1999.

tial operational deliveries in July 1999. As a result, the initial operational test and evaluation (IOT&E) phase of the program was divided into two parts. The first part of the restructured IOT&E program included only testing of subsonic launches from B-52s. These tests, which proved highly successful, permitted the authorization of LRIP in August 1998 and meant that initial operational deliveries to B-52 squadrons could take place early in 1999, three to five months ahead of schedule. The second part of the restructured IOT&E program aimed at flight testing the fin-locking mechanism that had already been designed and ground tested. Program officials claim that this restructuring has had little effect on the production schedule. The only significant consequence, they argue, is that the money that was going to be used to incorporate a small electronics upgrade in the WCMD kit had to be spent on the software and fin-lock fixes.[46]

The JASSM program is still in the early stages of R&D. GAO published a report on the JASSM program in 1996 which concluded that in the long run, the risk of cost growth and schedule slippage was high.[47] This conclusion was based on the view that the JASSM development schedule was too short to permit maturation of the high-risk technical areas on the program: automatic target recognition, autonomous guidance, and aircraft integration.

Beginning in 1997, a variety of factors—including concerns over the level of technical risk remaining in the program—led to a restructuring of the program schedule. The original program schedule envisioned a 24-month PDRR phase beginning in June 1996, followed by a 32-month EMD phase beginning in June 1998. The nominal target date for the authorization of LRIP was April 2001. Spurred by declining Navy interest in the program and by significant congressional funding cuts in late 1997, however,[48] the Air Force restructured the

---

[46] "First WCMD B-52 Test Prepared for Continued Development Testing," *Aerospace Daily*, April 29, 1998.

[47] U.S. General Accounting Office, *Precision-Guided Munitions: Acquisition Plans for the Joint Air-to-Surface Standoff Missile*.

[48] The Navy was convinced that the Boeing SLAM-ER, a modification of the existing Navy AGM-84 SLAM system (itself a modification of the Harpoon), would meet its requirements at less cost than the JASSM. Like JASSM, SLAM-ER is slated to have an automatic target acquisition system. It will have a more than 100-nm standoff range and will deliver a 500-lb. warhead. Congress authorized an analysis of which system best served both services' needs. A GAO study concluded that JASSM could potentially be

PDRR phase. First, it was decided to down-select to one contractor on April 1 rather than in late June or early July, at the planned beginning of EMD, in order to save money. Second, the PDRR phase was extended by about three months. Eventually this evolved into a six-month extension or more, until November 1998. Thus, if one counts from the original contract award to the two contractors for the PDRR phase (June 1996), the PDRR phase has been extended by 25 percent compared to original estimates.[49] This increase in the PDRR phase provided more time for the contractor and the Air Force to reduce technical risk prior to full-scale development. In addition, technical risk was further reduced through the elimination of some of the developmental tasks that had to be completed during PDRR. For example, given that the Navy had decreased its involvement in the program, the need to focus on the early integration of JASSM with the F/A-18E/F fighter was eliminated.

In November 1998, press accounts reported that the DoD also intended to restructure the full-scale-development EMD phase by lengthening it considerably. According to these accounts, the EMD phase would be stretched from 34 months (originally 32 months) to 40 months—an increase of 25 percent—in order to further reduce technical risk prior to flight testing. According to a program official, "We [the Air Force and Lockheed Martin] decided that we needed to do more ground and captive carry testing than what we planned to in order to not have big surprises during the flight test program."[50] These schedule increases, officials predicted, would cause a commensurate increase in overall R&D costs.

---

fielded earlier with superior capabilities and at less cost than the upgraded SLAM-ER Plus version, the development of which would be necessary to meet all key JASSM performance objectives. OSD directed the Navy to maintain at least minimal participation in the JASSM program, but with the withdrawal of the F/A-18E/F as a "threshold" aircraft, active Navy participation was essentially ended. See "GAO Finds No Reason to Terminate JASSM," *Aerospace Daily*, September 29, 1998.

[49]A contributing factor was that the program experienced many weeks of delay after the down-select to two contractors because of an official protest filed by Hughes, one of the contractors that had lost in the first phase of the program. Because of the heavy use of past performance criteria by the government, all three of the munitions programs examined here experienced formal protests after the initial down-select process, which led to considerable lost time and effort. In all cases, however, the government won its case against the protests.

[50]Terry Little, JASSM Program Director, quoted in "Approval of Extended JASSM EMD Program Seen Imminent," *Aerospace Daily*, November 11, 1998.

At the time, these schedule extensions did not appear to be the product of major technical difficulties or problems caused by the innovative commercial approach; rather, they seemed to be the result of a development schedule that the program director characterized as "unrealistic" given the level of technical risk involved. Even after the extension, the program director characterized it as "still the most aggressive new development for a weapon" in a long time.[51]

Then, in April 1999, the first JASSM flight test vehicle crashed, delaying the flight test program at least one month. A "makeup" flight test was scheduled for August. On the 12th of that month, Lockheed completed a successful separation and maneuver flight test.[52] Two weeks later, the Air Force announced a major restructuring of the EMD program. The Air Force and Lockheed agreed to delay the decision to begin LRIP by ten months, from January to November 2001, to permit additional flight tests of production-standard JASSM vehicles. The Air Force blamed technical problems and the contractor for the delays. According to press accounts, problems were being experienced with engine development, missile casing, and the air data system.[53]

It is unclear whether any of these developmental problems are related to the CMI approach adopted in the program. However, it is likely that they are normal events typically encountered in the development of any complex new system. Interestingly, JASSM's competitor, the SLAM-ER, failed its operational tests in August 1999, and as a result the Navy delayed full-scale production until at least the spring of 2000. SLAM-ER is usually considered a technologically lower-risk program than JASSM because it is not a new development but rather a modification of the Harpoon/SLAM series of missiles.

The original 56-month development program has now been extended to 78 months, an increase of nearly 40 percent. Nonetheless, the new schedule is still well below the average munitions develop-

---

[51]Ibid.
[52]See "JASSM Takes First Step in Flight Testing," *Jane's Defence News*, August 25, 1999.
[53]"USAF Postpones JASSM Decision Until 2001," *Aerospace Daily*, August 30, 1999.

ment schedule of 110 months, according to the JASSM program director.[54]

Assuming that the new development schedule can be met, JASSM will still be developed in less time than TSSAM. TSSAM was canceled after roughly eight years of R&D, and development still had not been completed. JASSM is now scheduled to be fully developed in six and one-half years from program initiation—a schedule improvement over TSSAM of at least 18 percent.

SUMMARY OF AIR FORCE AR MUNITIONS PROGRAMS COST SAVINGS

In summary, despite initial development schedules that were probably unrealistically aggressive, these three munitions programs appear to be largely meeting their cost and performance expectations within overall development periods that are far shorter than traditional programs for similar systems.

These data suggest that R&D savings in the range of 15 to 35 percent may be possible in programs that are fully restructured in a commercial-like manner in accordance with CAIV concepts. The likely scale of anticipated production savings is much more uncertain. However, the three best-documented cases—JDAM, WCMD, and JASSM—suggest that large savings of up to 65 percent are possible, at least in programs for relatively less complex systems with high production runs.

Some additional qualifications must be noted in discussing these outcomes. The reforms used on these pilot programs, for example, have not been widely used as an integrated package outside of these AR demonstration programs. Furthermore, these AR pilot programs are relatively small and are characterized by low technological risk, commercial derivative items, and very large production runs. Thus, the scale of potential cost benefits for a large, complex weapon system employing high-risk, cutting-edge technology remains uncertain. Finally and most significantly, several of these programs have only recently entered the LRIP stage, or have not completed EMD.

---

[54]"JASSM Schedule Slip Costs $53 Million," *Aerospace Daily*, August 31, 1999.

Having completed our detailed review of these three programs, we now turn to briefer summaries of other programs that share similar traits. Estimates of AR savings for these programs are also presented.

Chapter Five

OTHER COMMERCIAL-LIKE ACQUISITION REFORM PILOT PROGRAMS

INTRODUCTION

This chapter examines several commercial-like AR pilot-program case studies in much less detail than the preceding case studies of the three U.S. Air Force munitions programs. These include two Air Force space AR pilot programs, two additional DAPPs, and three unusual programs sponsored by DARPA. The purpose of this quick review was to serve as a "sanity check" on the program-structuring philosophy and cost savings estimates that we discovered in our more detailed case studies of the Air Force AR munitions pilot programs. We were also interested in determining if robust data existed to support claims of cost savings on widely different types of systems with different sponsoring government authorities.

If anything, this quick review of other pilot programs increases our concerns about the reliability and robustness of currently available cost savings projections for AR pilot programs. All of the qualifications and caveats made in earlier parts of this report apply with equal or greater relevance to the estimates of cost savings presented here.

Finally, this chapter also presents the subjective consensus views on AR pilot programs and potential savings from commercial-like program structuring that we garnered from our extensive interviews with industry officials.

SELECTED U.S. AIR FORCE SPACE AR LEAD PROGRAMS: SBIRS AND EELV

Space-Based Infrared System (SBIRS)

SBIRS is an Air Force program for the development and deployment of space-based surveillance systems for ballistic missile warning, defense, and intelligence. SBIRS is intended to replace the existing Defense Support Program (DSP) satellites.[1] The basic SBIRS system consists of two space elements: SBIRS High Component and Low Component. The High Component is the main element driving the development effort. Approved for development in October 1996, the SBIRS High Component includes four geosynchronous earth orbit (GEO) satellites, two highly elliptical earth orbit (HEO) satellites, and a consolidated ground processing station. SBIRS High meets the basic threshold requirements of the program. First delivery for the High Component was originally scheduled for FY02. The Low Component is intended to make use of 24 low-earth orbit (LEO) satellites. The Low Component provides unique precision midcourse tracking capabilities for ballistic missile defense (BMD). The Low Component was originally planned for flight demonstration in FY98 followed by a deployment decision in FY00.[2]

The need for SBIRS originally arose in 1995 after the cancellation of the Follow-on Early Warning System as a result of cost growth and technical problems. Technical aspects of replacing the DSP system necessitated a tight developmental schedule for SBIRS. This situation, combined with funding limitations, led OSD and the U.S. Air Force to designate SBIRS in its early stages as an AR lead program. As a result, interested contractors were provided with the ORD rather than with detailed technical specifications. The government required only two military standards. In addition, the traditional SOW was replaced with a high-level SOO.

Initial program cost goals were established during the 1994 Infrared Summer Study. At the time, government officials recognized that insufficient funds existed to meet all the potential user community re-

---

[1] First deployed in the early 1970s, DSP satellites use infrared sensors to detect ground and space missile launches and nuclear detonations.
[2] See U.S. Air Force, *Space-Based Infrared System Fact Sheet*, no date.

quirements. As a result, the program was subjected to an intensive CAIV process that included an aggressive must-cost total program cost objective. The government incentivized the CAIV process by promoting contractor competition and making R&D, procurement, and life cycle costs (LCCs) key criteria for down-selection.

In August 1995, two contractor teams, one led by Lockheed Martin and the other led by Hughes (now Raytheon), each received $80 million 15-month contracts for a preengineering, management, and development phase of SBIRS High. Contractors were encouraged to exploit commercial and off-the-shelf technology to reduce R&D, procurement, and LCCs. More significantly, they were given nearly total design and configuration control through the Total System Performance Responsibility (TSPR) concept and approached the problem in a cohesive way through IPTs in order to come up with innovative solutions that met performance requirements as well as cost target goals. Both contractors responded with price bids below the government's CAIV estimate of most likely cost. The government eventually awarded Lockheed Martin the $1.8 billion ten-year High Component EMD contract in October 1996. Total program costs for the High Component have been estimated at more than $7.6 billion (TY dollars).[3]

The government encouraged Lockheed to continue its cost/performance trade studies and work to further reduce costs. As an incentive, the government approved an unusually large 20 percent award fee in the contract, half of which depended on successful cost management. To receive the highest rating for this area, the contractor had to further reduce the procurement price without sacrificing critical performance capabilities. Lockheed established a goal of a 10 percent reduction in the cost of the first three satellites compared to its Best and Final Offer proposal during the competition. As an incentive to maintain performance capabilities, reliability, low LCC, and schedule, Lockheed Martin committed to paying up to 8 percent of the contract value if it failed to meet critical mission and program milestones.

---

[3]Department of Defense, Office of the Director, Operational Test and Evaluation, *DOT&E FY98 Annual Report.*

Meanwhile, the competitive Flight Demonstration System (FDS) program and the Low Altitude Demonstration System (LADS) program for the SBIRS Low Component moved forward with the two main teams of Lockheed/Boeing and TRW/Raytheon. Later, a third team led by Spectrum Astro and including Northrop Grumman joined the SBIRS Low competition. Spectrum Astro is a relatively new entrant to the military satellite sector and has been one of the pioneers of NASA's low-cost COTS-based space vehicles. Its involvement in SBIRS Low symbolized DoD attempts to bring new, more commercially oriented firms into the defense acquisition process. Other AR measures during this period included Lockheed's decision to base its SBIRS LADS on a modified commercial LM700 satellite bus. This decision was made to achieve cost savings by leveraging Lockheed's commercial satellite production facilities. The LM700A bus is used by the Iridium commercial telecommunications satellite program. More than 80 LM700A buses have been manufactured at an innovative "mass production" commercial facility. According to Lockheed, leveraging the commercial production line "lowers our costs substantially."[4]

During the 1994 Infrared Summer Study, officials estimated that SBIRS R&D costs would be approximately 15 percent below the likely R&D program costs had a traditional acquisition approach been used.[5] This estimate, shown in Table 5.1, remained officially valid at least through 1997. However, funding cuts, a tight schedule, and persistent technical problems led government officials in 1998 to reassess the progress of both the SBIRS High and SBIRS Low components. In early 1999, the government canceled both SBIRS Low demonstration programs owing to an anticipated 57 percent cost growth ($900 million versus $575 million) on the FDS program and a 62 percent cost growth on the LADS program ($240 million versus $149 million).[6] During 1999, officials also announced that cost growth was expected on the SBIRS High R&D program. Pentagon of-

---

[4]Lockheed Martin press release, 1998.

[5]Thomas E. Rosensteel, *An Implementation of Cost as an Independent Variable (CAIV: The Space Based Infrared Systems (SBIRS) High Component Program*, SBIRS Program Office, no date.

[6]Testimony of Acting Air Force Secretary F. Whitten Peters as reported in "SBIRS Low Teams Put Finishing Touches on Revised Proposals," *Aerospace Daily*, April 6, 1999.

ficials decided to restructure SBIRS by spending more money on risk reduction and technology demonstration programs, especially for SBIRS Low, and by stretching out the development schedules for both the High and Low components. As of mid-1999, the SBIRS schedule was expected to slip at least two years.[7]

Part of the restructuring of SBIRS Low included the down-selection to two contractor teams for the final three-year PDRR phase. In August 1999, the Air Force selected the TRW and Spectrum Astro teams to continue SBIRS Low work, awarding both teams fixed-price contracts worth $275 million each.

Table 5.1

Selected Air Force Space AR Pilot Program Estimated Savings

	R&D (%)	Production (%)	Estimate Source	Quality
EELV	20–33	25–50 (30+)[a]	1, 2, 3	Analysis
SBIRS[b]	15		4, 5, 6	Analysis

SOURCES:
(1) U.S. General Accounting Office, *Evolved Expendable Launch Vehicle: DoD Guidance Needed to Protect Government's Interest*, GAO/NSIAD-98-151, June 1998.
(2) Samuel A. Greaves, *The Evolved Expendable Launch Vehicle (EELV) Acquisition and Combat Capability*, Air Command and Staff College, Maxwell AFB, AL, March 1997.
(3) EELV Program Office Fact Sheets, October 1998.
(4) Statement of the Under Secretary of Defense for Acquisition and Technology Honorable Paul G. Kaminski Before the Acquisition and Technology Subcommittee of the Senate Committee on Armed Services on Defense Acquisition Reform, March 19, 1997.
(5) Thomas E. Rosensteel, *An Implementation of Cost as an Independent Variable (CAIV): The Space Based Infrared Systems (SBIRS) High Component Program*, SBIRS Program Office, no date.
(6) Jay Moody, *Achieving Affordable Operational Requirements on the Space Based Infrared System (SBIRS) Program: A Model for Warfighter and Acquisition Success?* Air Command and Staff College, Maxwell AFB, AL, March 1997.
[a]After program restructuring in 1999 employing commercial-like contracts for launch services, Air Force officials estimated a savings to the government in launch costs through 2020 of more than 30 percent compared to using existing expendable launch vehicles.
[b]SBIRS High Component program only.

---

[7]The SBIRS High first launch was delayed by two years from FY02 to FY04. The SBIRS Low first launch was also delayed two years to FY06.

SBIRS Low is planned to be the largest U.S. government satellite constellation ever built. The government has applied a must-cost goal of $79 million per satellite. This is an ambitious goal, since the cheapest satellite of this type built in the past cost on the order of $150 million. TRW's approach to the cost goal is based on using COTS-based hardware and software elements to lower costs and raise capabilities. TRW's focus will be on COTS-based software elements, because historically software has accounted for some 25 percent of costs compared to less than 10 percent for satellite hardware.[8] TRW plans to use space-qualified, radiation-hardened PowerPC (RHPPC) 603e modules and support chip modules. These modules are expected to cost on the order of $10 million compared to specially developed space-unique modules such as the RH32, which cost on the order of $25 million. The approach will be to maximize board-level COTS commonalities using COTS-like architectures to reduce costs.[9]

Given the technical risks involved, the cost growth and schedule restructuring experienced by SBIRS to date seems to have resulted primarily from an overly aggressive schedule. Budget cuts and budget restructuring have also caused problems. It is unclear whether the innovative AR approach adopted will ultimately result in the 15 percent R&D savings originally anticipated. Both SBIRS High and SBIRS Low are still in their earliest stages of development. It is possible, however, that AR will ultimately contribute to lower costs through COTS insertion as well as to fewer problems with cost and schedule than might otherwise have been experienced.[10]

EELV (Evolved Expendable Launch Vehicle)

Like the Tier III– DarkStar and Tier II+ Global Hawk programs, the EELV program is operated under Section 845 "Other Transactions"

---

[8] The rest of the program costs are primarily for launch services.

[9] This paragraph is based on a briefing by Eric Johnson and Dean Brenner, Honeywell Space Systems, presented to the COTSCon Conference held in San Diego in December 1999.

[10] According to the Department of Defense, Director, Operational Test and Evaluation, *DOT&E FY98 Annual Report*, SBIRS suffers from "a compressed schedule" with "zero schedule margin" to compensate for a high-risk program with extremely demanding software development and integration challenges.

Authority (OTA).[11] Section 845 OTA eliminated nearly all normal procurement statutes and Federal Acquisition Regulations (FARs) to permit maximum program flexibility in developing demonstration prototypes of weapon systems.[12] Originally used primarily by DARPA, OTA is now being more widely applied to large service-run developmental programs such as EELV. Indeed, EELV is by far the largest R&D and procurement program ever conducted under Section 845 OTA.

The U.S. Air Force designated EELV as a lead program for AR and as a CAIV flagship program. The government established a goal of 50 percent cost reduction and a cost reduction minimum threshold of 25 percent in recurring launch costs compared to current Delta and Atlas launches. Like other CAIV programs, EELV established KPPs that had to be met but permitted contractors to meet these KPPs any way they thought best by using their own design and technical engineering ingenuity. The four KPPs are:

- Mass to orbit;
- Design reliability;
- Standard launch pads; and
- Standard payload interface.

All other requirements are tradable, particularly against cost.

Four contractors (Alliant Techsystems, Boeing, Lockheed Martin, and McDonnell Douglas) were each awarded $30 million contracts to compete in the initial low-cost concept validation (LCCV) phase of the EELV program. When this phase ended in November 1996, the Air Force down-selected to two contractors—Boeing and Lockheed—to begin the preengineering and manufacturing development phase. This phase lasted 17 months. Each contractor received $60 million.

The DoD originally planned to down-select to one contractor at the end of this phase. The winning contractor would be awarded a cost

---

[11]Section 845 of the National Defense Authorization Act for Fiscal Year 1994 (P.L. 103-160, November 30, 1994).
[12]See Department of Defense, *Defense Acquisition Handbook*.

plus award fee (CPAF) contract of approximately $1.5 billion and would proceed with EMD. In November 1997, the DoD approved a revised acquisition strategy based on AR principles and on dual-use CMI. Projecting a much larger commercial market for the EELV launchers than originally anticipated, the Air Force decided to maintain competition between the two contractors throughout EMD and production. The government contribution to R&D would be capped at $500 million per contractor. Each contractor was expected to invest between $800 million and $1.3 billion of its money into R&D. This is because the government estimated that some two-thirds of the launch market available to the contractors would arise from commercial sources. In addition, the government would pay for fewer test flights. Finally, the government decided to contract for launch services through 2020 in a commercial manner after EMD was complete.

EELV entered the equivalent of EMD in October 1998. The government signed two $500 million development contracts through 2002 with Boeing and Lockheed. In addition, Boeing was awarded a $1.38 billion initial launch services (ILS) contract to launch 19 government payloads between 2002 and 2006, while Lockheed received an ILS contract for $650 million for the launch of up to nine government payloads. In short, the government decided to partially subsidize EMD for the systems proposed by both companies and then contract with both for launch services in a commercial manner. Government officials reasoned that the EMD subsidies would permit the development of both EELV systems, thereby providing significant competition for future launch services—resulting in lower costs for both government and commercial customers. The two prime contractors continued to retain almost total configuration control and responsibility for price, cost, and technical management. The only government requirements were posed in terms of performance: (1) launch reliability, (2) standardization of payload carrier, (3) dual capability for military and commercial payloads, and (4) standardization of launch pad. Government program officials calculated that this commercial approach would save the government more than 30 percent in launch costs through 2020 compared to the use of existing Atlas, Delta, and Titan rockets.[13]

GAO, however, calculated that while the government was now committed to an EMD investment of $500 million less than the original approach (hence the 33 percent savings on EMD), this did not take into account reimbursement to the contractors for independent research and development (IR&D).[14] Moreover, because the plan changed to support two contractors rather than one, some of the $500 million in EMD savings would be diverted to the support of two additional government-unique test launches. Therefore, GAO estimated that the EMD savings would be in the range of 20 percent. GAO also estimated that the recurring production and launch cost savings could be as high as 37 percent, but a variety of unknowns and uncertainties made precise estimates of savings impossible. Nonetheless, the U.S. Air Force adheres to its original estimates of both EMD and production/launch savings. These estimates are shown in Table 5.1.

In early 2000, press accounts reported that an audit conducted by the DoD Inspector General's office had concluded that one of the EELV contractors, Lockheed Martin, had received some $103 million for EMD above the fixed $500 million level through IR&D reimbursement. The audit allegedly criticizes the Air Force for having insufficient insight into contractor EELV expenditures, inferring that this was due in part to the waiver of traditional cost oversight mechanisms. The Air Force rejected this criticism, noting that contractors make the decision on how to allocate IR&D monies.[15]

---

[13]See, for example, Tom Kuhn, "EELV: A New Rocket for the Millennium," *Airman*, March 1999.

[14]U.S. General Accounting Office, *Evolved Expendable Launch Vehicle: DoD Guidance Needed to Protect Government's Interest*, GAO/NSIAD-98-151, June 1998.

[15]Tony Capaccio, "Lockheed Besting Boeing in Funds," *Denver Post*, March 3, 2000.

SELECTED DEFENSE ACQUISITION PILOT PROGRAMS (DAPPs)

One DAPP—JDAM—has already been discussed at great length above. This subsection examines two additional DAPPs: FSCATT and JPATS. Two other DAPPs, the CDE and C-130J, are not discussed in detail here because the information we received on them was mostly proprietary and competition sensitive. The final DAPP—the Defense Personnel Support Center (DPSC)—is not examined in this report because it is not directly related to a specific weapon system.

Fire Support Combined Arms Tactical Trainer (FSCATT)

The FSCATT program calls for the development of a training simulation system in two phases.[16] The first phase aims at developing a system that can train the Army Artillery Gunnery Team without the need for live firing exercises. Battery-level training will be provided by exercising and measuring the performance of individuals, gun crews, and battery teams. The second Phase of FSCATT aims at developing a broader simulation that integrates artillery training into the entire spectrum of the Combined Arms Tactical Training environment.

The DoD incorporated DAPP principles into the initial Phase I RFP for FSCATT when it was sent out in May 1994. A major focus of the FSCATT program was to promote the participation of commercial firms in defense R&D and to demonstrate the incorporation of dual-use and existing off-the-shelf technology into a defense product. Eighty-one potential bidders received the RFP. In June 1995, Hughes Training Inc. (HTI) won the FSCATT contract.

Between October and December 1995, HTI and the program office conducted a "would cost estimate" exercise to verify the anticipated AR cost savings for FSCATT. DCAA then verified and certified the results of this exercise. The results showed a program savings of 13.5

---

[16]This description is drawn from Office of the Deputy Under Secretary of Defense, Acquisition Reform, Pilot Program Consulting Group, *PPCG 1997 Compendium of Pilot Program Reports,* Office of the Deputy Under Secretary of Defense (Acquisition Reform), and Department of Defense, Pilot Program Consulting Group, *Celebrating Success: Forging the Future.*

percent, or $14 million. The breakout of the savings is shown in Table 5.2.[17]

The HTI estimate for the cost of FSCATT R&D alone under a traditional program versus a DAPP projected an AR R&D savings of 16 percent. Another source claimed that the final analyses projected R&D savings to be in the range of 34 percent and production savings to be 7 percent.[18] These estimates are summarized in Table 5.3. In early 1999, the Deputy Under Secretary of Defense, Acquisition Reform, reported an overall 13.5 percent contract cost savings on FSCATT due to AR.[19]

Table 5.2

FSCATT: Breakdown of Estimated AR Program Savings

	Percent of Total Savings	Percent of Program Savings
Regulatory burden	14.8	2.0
Quality assurance	1.5	0.2
Data/configuration management	0.7	0.1
Program management	2.2	0.3
Test/evaluation	0.7	0.1
Contract	9.6	1.3
Commercial insertion/program structure	85.2	11.5
Design/assembly	7.4	1.0
Software	11.1	1.5
Manufacturing	34.1	4.6
Parts/procurement	25.2	3.4
RAM/ILS[a]	7.4	1.0
Total	100%	13.5%

[a]Reliability, Availability, Maintainability/Integrated Logistics Support.

---

[17]Original categories were organized into two major categories by the authors.

[18]*Army Acquisition Reform Newsletter*, Issue 6, January 1996, and presentation by Michael T. Smith, Chairman, Hughes Aircraft Company, and Vice Chairman, Hughes Electronics Corporation, Army Acquisition Roadshow Symposium, Orlando, Florida, April 9, 1996.

[19]Statement of Stan Z. Soloway, Deputy Under Secretary of Defense, Acquisition Reform, March 2, 1999.

Table 5.3

Projected FSCATT AR Savings (Multiple Sources) (%)

Total Contract	R&D	Production
13.5	16–34	7

Joint Primary Aircraft Training System (JPATS)

JPATS is one of the original DoD-designated DAPPs. It aims at producing a new primary training aircraft for the Air Force and the Navy to replace the USAF T-37B and the Navy T-34C.

An extensive source selection process for JPATS commenced in May 1994. The DoD evaluated candidate aircraft from seven contractors. Most of the candidates were existing aircraft that had already been partially or completely developed by foreign contractors but were planned for manufacture by a U.S. contractor. In June 1995, the DoD selected Raytheon Beech Aircraft's JPATS proposal based on the Pilatus PC-9 aircraft. Although the DoD estimated roughly $7 billion for the manufacturing development (MD), production, and initial support of more than 700 aircraft, its designation of JPATS as a DAPP was expected to result in a far less expensive program. Protests from Cessna and Rockwell prevented the award of the JPATS contract until February 1996. At that time, Raytheon received a contract for MD, initial production, development of the Ground-Based Training System (GBTS), and support. Because of AR, the structuring of the program in a more commercial-like manner, and other factors, total program value at that time was projected to be approximately $4 billion. Program attributes included partial long-term contractor logistics support and competitive commercial subcontracting and development of the GBTS.[20]

The original JPATS acquisition strategy aimed at procuring an existing aircraft as a nondevelopmental item (NDI) with minimal modification. As the program evolved, however, various service requirements resulted in the need to make substantial modifications to the baseline Pilatus aircraft during MD. These changes included a new canopy, a new "zero-zero" ejection seat, new instrumentation and

---

[20]See "JPATS Contract Awarded," *Air Force News*, June 1995, and "Beech Aircraft Gets Trainer Contract," *Air Force News*, February 1996.

avionics, a redesigned tail and nose, a different engine, and a redesigned hydraulic system, wing structure, and avionics bay. As a result, the MD phase cost more than the initial FY92 Program Office estimate (POE), as shown in Table 5.4. However, the FY92 estimate is not directly comparable to the others because it represents a "generic" JPATS rather than a specific design proposal. The other two estimates are based on the winning Beech (now Raytheon Beech) JPATS aircraft.

The FY97 POE shows considerable MD savings over the FY95 POE. In addition, both FY95 and FY97 POEs show savings over the original FY92 Program Office low estimate for production, O&S, and total program costs. However, total production costs for 711 aircraft increased in the FY97 POE compared to the FY95 POE. Again, none of these estimates are exactly comparable because various aspects of the program have changed between each estimate.

Table 5.5 shows the percentage savings in the FY97 POE compared to FY95 and FY92. Although the FY97 POE for MD is much higher than the FY92 POE, this is attributable in part to the dramatic increase in modifications to the baseline aircraft required by the services. As noted above, the FY92 estimate is a generic estimate based on analogies to the T-1, T-46, market surveys, and other historical factors and is not truly representative of the actual JPATS design winner. The FY97 MD estimate shows a 13.6 percent decline compared to the FY95 estimate. Although production costs in the FY97 POE are larger

Table 5.4

JPATS Program Cost Estimates (Base Year [BY] 1995 $M)[a]

	Manufacturing Development	Production	O&S	Total
FY92 POE (low)	139.14	4,094.6	15,297.1	19,530.8
FY95 POE	304.9	2,213.9	12,144.9	14,663.7
FY97 POE	263.4	2,802.1	8,820.9	11,886.4

[a]Based on data from Office of the Deputy Under Secretary of Defense, Acquisition Reform, Pilot Program Consulting Group, *PPCG 1997 Compendium of Pilot Program Reports*.

Table 5.5

FY97 POE Projected JPATS and Other Savings as Percentage of FY92 and FY95 POEs

	R&D	Production	Operations and Support (O&S)	Program Total
FY97 vs. FY92 POE	(89% cost growth)	31.6	42.3	39
FY97 vs. FY95 POE	13.6	(26.6% cost growth)	27.4	18.9

than the FY95 POE, total program costs have declined by nearly 19 percent, primarily because of O&S savings. In early 1999, the Deputy Under Secretary of Defense, Acquisition Reform, testified that the JPATS program had achieved an estimated 49 percent reduction in overall contract cost.[21] According to program officials and the contractor, a significant but nonquantifiable percentage of these savings can be attributed to AR.

SELECTED DARPA SECTION 845 "OTHER TRANSACTION" PROGRAMS

DARPA High-Altitude Endurance Unpiloted Aerial Vehicle and Arsenal Ship Programs

This category is made up of three programs initiated by DARPA for the development and possible production of three major platforms: two high-altitude endurance (HAE) unpiloted air vehicles (UAVs) and a program for a new class of naval surface combatants called Arsenal Ships (which has since been canceled). The information on these programs was drawn from other ongoing RAND research and from interviews with contractors.[22] The UAV program originally included

---

[21] Statement of Stan Z. Soloway, Deputy Under Secretary of Defense, Acquisition Reform, March 2, 1999. It is unclear in this document what precisely is meant by the term "contract cost."

[22] This research is currently being conducted by Jeffrey Drezner, Robert Leonard, and Geoffrey Sommer. See also earlier RAND research published on this topic in G. Sommer, G. K. Smith, J. L. Birkler, and J. R. Chiesa, *The Global Hawk Unmanned Aerial Vehicle Acquisition Process: A Summary of Phase 1 Experience*, Santa Monica: RAND, MR-809-DARPA, 1997.

the Tier II+ Global Hawk under development by Teledyne Ryan Aerospace (TRA) and the stealthy Tier III– DarkStar formerly under development by Lockheed Martin and Boeing. The DoD canceled DarkStar in February 1999, reportedly because of anticipated operational performance difficulties with the air vehicle and recognition of the Air Force's inability to provide sufficient funding to support the acquisition of both DarkStar and Global Hawk. Despite its problems, the DarkStar program is still worth including here because of some of its interesting AR features.

Both of these programs were designated as Advanced Concept Technology Demonstration (ACTD) programs and were operated under the auspices of DARPA's Section 845 OTA. The programs contained novel characteristics similar to formal DAPPs. For example, both UAV programs had only one formal requirement: a hard must-cost unit flyaway price (UFP) of $10 million (FY94); all other aspects of the programs were flexible and could be traded off against cost.[23]

In the case of the now-canceled DARPA/Navy Arsenal Ship Program (ASP) for the development of a revolutionary new class of ships, the competing contractors shared some 50 percent of the costs during the concept development and design phases (Phases I and II) and anticipated paying for a significant portion of the EMD prototype phase (Phase III) prior to program cancellation. Like the UAV programs, the ASP operated under Section 845 OTA and exhibited many AR characteristics similar to those of other programs discussed here. For example, the principal government requirement was a hard "must not exceed" unit sailaway price (USP) of $550 million and a target USP of $450 million.

The DarkStar and later versions of the Global Hawk contractor agreements included terms that required the contractors to share in unanticipated R&D cost growth. The DarkStar Phase II (prototype) R&D baseline agreement was essentially a traditional cost plus fixed fee/incentive fee (CPFF/IF) instrument. The government agreed to pay all Phase II R&D costs up to $115.7 million. The contractor could earn a relatively small fixed fee as well as a small incentive fee for

---

[23]UFP is defined as the average price of air vehicles 11–20, including sensor payload, for both programs.

meeting performance goals in several areas.[24] These fees could amount to some $8 million to $9 million, or roughly 8 percent of R&D cost.

In a radical departure from traditional programs, however, the DarkStar contractors agreed to pay 30 percent of Phase II R&D costs if they rose above $115.7 million and 50 percent of R&D costs above $162 million. Finally, the parties agreed to an absolute cap of $220 million on Phase II. Since relatively serious technical problems were encountered during the prototype flight test program, resulting in a lengthening of the Phase II schedule, it is likely that the $220 million ceiling was or would have been reached had the program not been canceled.[25] Had the ceiling been reached, the contractor would have been responsible for paying for nearly $43 million, or more than 40 percent of a cost overrun of $104 million. According to the contractor, this has been a painful experience and would have been a strong incentive to reduce technical risks and control costs in future phases of the program. It has saved the government a significant amount of money and made the contractor a risk-sharing partner in the development program, as is common in the commercial world.[26]

Although the Global Hawk program also experienced technical problems, cost growth, and schedule slippage during Phase II R&D, it has continued as a viable program. Program managers had originally planned to impose cost and performance discipline on the program by maintaining competition with at least two contractors throughout Phase II. However, funding shortfalls required an early down-select to one contractor. The Phase II agreement remained a traditional CPFF/IF instrument. As a result of significant cost growth, the parties renegotiated the Phase II agreement in mid-1997. The new

---

[24]These areas covered performance of the air vehicle (altitude and endurance), sensors (radar, electro-optical, IIR), and the command-and-control ground station.

[25]In principle, ACTDs are intended to allow DARPA, in close association with potential user services, to rapidly integrate relatively mature technologies into prototypes to demonstrate a useful operational capability. However, the development of both DarkStar and Global Hawk entailed considerable technological risks, particularly in the areas of systems integration and stealth.

[26]Whether or not this results in true cost savings for the government depends on how the contractors treat these added expenses. It is possible that the extra cost could be passed back by the contractor to the government through higher overhead rates or during production.

agreement resembled the DarkStar Phase II clauses that required the prime contractor to pay a percentage of cost overruns beyond a certain threshold, and caps total government expenditure on the phase. The new Global Hawk program also required that major subcontractors share in cost overruns, as is now common in the commercial aircraft industry.[27]

The costs of the first Global Hawk flight vehicles have reportedly grown dramatically over original estimates. As of this writing, it is not known whether the original must-cost production cost goals will be met. Most observers speculate that Global Hawk will cost considerably more than originally anticipated, partly because of requirements and design changes implemented by the Air Force after it took over the program.[28]

The Navy canceled the ASP in October 1997, near the end of the concept and design development phase (Phase II) as a result of congressional reductions in program funding. Enough data had been generated, however, to permit a comparison of the projected Arsenal Ship USP with the original Navy estimate of USP if the Arsenal Ship had been developed in a traditional manner. RAND analysis suggests that the AR measures adopted would have resulted in a USP savings of at least 30 percent compared to a traditional program. These data and estimated cost savings on the other two DARPA OTA programs are shown in Table 5.6.

---

[27] See Footnote 24.

[28] According to one source, unit price has escalated from $10 million to $15 million. See Karl Schwarz, "Global Hawk Convinces U.S. Air Force," *Flug Revue Online*, November 2000.

Table 5.6
DARPA Section 845 OTA HAE UAV R&D Programs

Program	R&D Cost Savings (%)	Source	Estimate Quality
Tier III– DarkStar	20[a]	1, 2	Forecast
Tier II+ Global Hawk	3[b]	2	Analysis
Arsenal Ship	30[c]	3	Analysis

SOURCES:
(1) Interviews conducted by the author with Lockheed Martin officials.
(2) RAND research in progress conducted by Jeffrey Drezner, Robert Leonard, and Geoffrey Sommer. See Jeffrey A. Drezner and Geoffrey Sommer, *Innovative Management in the DARPA HAE UAV Program: Phase II Experience*, Santa Monica: RAND, MR-1054-DARPA, December 1998. See also earlier RAND research published on this topic in G. Sommer, G. K. Smith, J. L. Birkler, and J. R. Chiesa, *The Global Hawk Unmanned Aerial Vehicle Acquisition Process: A Summary of Phase 1 Experience*, Santa Monica: RAND, MR-809-DARPA, 1997.
(3) RAND research in progress conducted by Jeffrey Drezner, Robert Leonard, and Geoffrey Sommer. See R. S. Leonard, J. A. Drezner, and G. Sommer, *The Arsenal Ship Acquisition Process Experience: Contrasting and Common Impressions from the Contractor Teams and Joint Program Office*, Santa Monica: RAND, MR-1030-DARPA, October 1998.

[a]Savings compared to total R&D program costs, which include a significant cost overrun. See main text for explanation.

[b]Savings compared to the total R&D Phase II agreement anticipated ceiling costs (which include a significant cost overrun) in accordance with the terms of the program agreement as amended August 4, 1997.

[c]USP production price.

ADDITIONAL OBSERVATIONS FROM INDUSTRY AND GOVERNMENT INTERVIEWS

The list below summarizes additional general observations made during RAND interviews with industry and government representatives about the commercial-style program structures previously discussed:[29]

1. Requirements reform (performance-based specifications) and CAIV are crucial for cost savings. CAIV essentially entails a trade-off of technical capabilities against cost. The key to CAIV is

---

[29]See Chapter One for a list of industry sites visited.

avoiding overdesigning and retaining only mission-essential capabilities.

2. CMI (commercial parts and technology insertion) has a high AR savings potential, especially in electronics.
3. Requirements reform, regulatory reform, CAIV, and especially contractor configuration control are all necessary to implement CMI.
4. Commercial-style programs with greater contractor cost sharing would be encouraged by reducing the constraints on foreign sales and technology transfer.
5. Commercial-like must-cost pricing goals combined with competition incentivize contractors to control costs.
6. Commercial-style R&D and production programs with contractor configuration control may require contractor logistics support once systems are fielded.
7. True dual-use (commercial and military) utilization of production facilities on a system or major-subsystem level is still rare. Government regulations and technology differences remain barriers.
8. The level of AR actually implemented on some pilot programs has been less than some contractors expected.

Requirements Reform, CAIV, and CMI

Nearly all contractors involved in AR programs stressed the importance of performance-based specifications, CAIV, and CMI. It appeared that the most significant cost savers in acquisition pilot programs were the elimination of capabilities not considered necessary or cost-effective for the performance of the mission (CAIV) and the use of commercial parts, processes, technologies, and production lines. CAIV was most important on the system level, whereas CMI seemed most applicable in the electronics area, especially on the parts level. Several examples from interviews are discussed below to illustrate some of the points listed above.

The JSF program is advanced by both contractors and the DoD as a flagship program for the use of CAIV. This program is being portrayed as a classic must-cost commercial-like effort. The only absolute requirement is the average unit recurring flyaway (URF) price. Moreover, there is no fixed ORD; rather, the definitive ORD will be the result of years of CAIV analysis and trade studies. In addition, there are only 12 requirements with numbers attached, and these are basic or "threshold" requirements such as mission radius and payload parameters.[30]

The two competing JSF prime contractors that remain are conducting numerous cost/operational-performance trade-off studies using three-dimensional design tools. Their goal is to provide the Joint Program Office with estimates of the sensitivities of each performance parameter for cost with the ultimate objective of identifying the cost drivers in the performance requirements and finding the "knee in the curve" in cost/performance trade-offs. Military users can then decide if the incremental change in performance in a given technical or operational performance area is necessary for achieving the overall mission and worth the incremental change in cost. Each major CAIV iteration can result—and has resulted—in a requirements reduction in order to meet the must-cost price goal. Thus, parameters such as mission radius, sustained g turn capability, wing area, and aircraft size and weight can be relaxed to reduce cost.

Like the JASSM program, both JSF contractors are using the same theater combat effectiveness model. The contractors and the government program officials maintain a dialogue on model assumptions and their effects on costs of specific performance parameters.

Some industry managers argued, however, that to have a truly commercial-like fighter development program, restrictions on foreign sales should be relaxed. These managers noted that many new commercial aircraft today are developed and manufactured by a risk-sharing strategic partnership of firms, often including foreign companies. Prime contractors should have a greater opportunity, it was argued, to form strategic relationships with foreign firms in order to

---

[30]It is possible that during R&D, contractors and the government might conclude that a specific threshold requirement cannot be achieved at the overall must-cost target URF price. In such cases, the government could lower the threshold requirement, raise the must-cost target URF price, or cancel the program.

encourage them to agree to contribute resources to nonrecurring engineering or to guarantee market access.

The restructuring of the C-17 program that took place in the mid-1990s is also an example of CAIV and a commercial-like program. One aspect of that restructuring is presented here.

The original C-17 design had a highly capable but costly engine nacelle design. The original nacelle design was developed solely to meet demanding performance requirements, not to control costs or to enhance manufacturability. Nacelle/engine performance requirements included the capability to back the aircraft up a 5 percent incline, full thrust reverser capability, and provision for safely unloading the aircraft while the engines were running. As part of the effort to meet these demanding performance requirements, the nacelle and engine designers developed a costly, large, and complex single-piece titanium casting for the tail cone assembly. A complicated and costly thrust reverser was also developed.

The overall cost of the C-17 eventually rose to an unacceptable level. In response, the Air Force and the contractor negotiated a new commercial-style fixed-price production contract. This contract led the prime contractor to impose new, much lower must-cost goals on all subassemblies and subsystems of the aircraft. U.S. Air Force and contractor teams reexamined the aircraft design and other aspects of the program to develop ways to reduce costs, including those of the complex nacelle and thrust reverser designs. The government funded a program to redesign the thrust reversers, which led to the adoption and incorporation of a much less expensive commercial-like thrust reverser. The production cost savings from using the commercial-like design are estimated by the contractor to be three times the cost of the government-funded redesign effort.[31] In addition, the titanium tail cone was eliminated and replaced by a more commercial-like design incorporating much more aluminum, thereby significantly reducing costs. Overall, the must-cost price reduction goal for the nacelle and some other substructures—dictated by the new fixed-price commercial-style production contract—is pegged at 27 percent, and it would appear that this goal is being met.

---

[31]It should be noted that in the commercial world, private companies generally pay all nonrecurring costs for the redesign of existing products.

The contractor claims that the redesign of the nacelles and thrust reversers "sacrificed zero capability." It is not clear whether this claim applies to the overall mission capability of the aircraft or to the specific technical performance capabilities of the engine nacelles and thrust reversers. Apparently, however, the use of commercial-style contracting and commercial technologies and design approaches has and will continue to significantly cut recurring unit production costs.

Avionics. In the area of avionics, some disagreement existed among contractors as to the use of commercial-grade electronics parts, but on the whole this approach was recommended as a means of substantially reducing avionics costs. However, only a few contractors provided specific cost savings estimates based on actual data and experience.

One contractor argued that mil spec parts are a major avionics cost driver, especially in the area of interface standards. The military has traditionally used custom interfaces, as in the case of the F-22; such custom interfaces determined the F-22 avionics architecture. Yet custom interfaces make upgrades with new-technology commercial chips difficult. COTS interface components and standards can be used to minimize nonrecurring and recurring costs.

Most contractors argued that commercial parts can save money while offering the same reliability, yield, and performance as mil spec parts if properly screened or used in more carefully controlled environments. Many automotive and industrial-grade electronic parts are close to the mil spec operating-temperature requirements of –55°C to +125°C. One problem is that virtually all new devices are commercial grade and are thus limited to an ambient temperature range of 0°C to 70°C. According to one manager, there are no mainstream processors that are industrial or automotive grade; automotive processors currently make up only 2 percent of the processors marketed. Up-screening commercial to industrial grade produces roughly an eightfold increase in cost. Moreover, commercial parts often will not work out of range, and those that do are not guaranteed by the manufacturer.

According to several contractors, the solution calls for a typical CAIV-type trade-off. Roughly 80 percent of commercial parts can meet radar mil spec high-temperature requirements in storage. The problem arises in the low end of the temperature range. One solution is to warm up the radars more slowly before beginning a mission. Since radars are liquid-cooled, the same system can be used to heat the system. Accepting a warm-up period two minutes longer, according to one contractor, allows a much larger number of commercial parts to be used, thus lowering costs. However, the military user must decide whether this is an acceptable cost/performance trade-off. In some cases, as with carrier-based aircraft, this may not be an operationally acceptable trade-off.

Implementation. At least one prime contractor on a pilot program complained that "there had not been nearly the magnitude of reform expected." Early drafts of the RFP, for example, contained mil specs and hard technical requirements. Later versions retained them as references, which looked just like wordsmithing to the contractors. Mil specs were also referenced in drawings. Furthermore, one contractor claimed, even the FAA essentially uses mil specs. Since several firms were in an intense competition, contractors stuck largely with mil specs in efforts to be conservative and to protect their position. Cost data were required as well, although the original proposal was not audited by DCAA. Although a CAS exemption was applied, the contractors supplied data to CAS standard because the government was intending to award a fixed-price incentive fee (FPIF) contract. In effect, most traditional CDRLs remained in place, at least from the contractor's perspective. Although an MMAS waiver was granted, an MRB was still required. Moreover, this capability had to be retained for other DoD programs. This contractor noted that while IPTs can help some things, they can also lead to increased rather than reduced oversight.

One contractor attempted to integrate its pilot program with its commercial production lines. This resulted in serious disruptions for commercial programs because of unanticipated delays and changes to the military program. On the other hand, the problems this contractor experienced, according to one manager, were not entirely attributable to government actions or inaction but were also caused by corporate inertia and unwillingness to change old ways.

SUMMARY OBSERVATIONS ON COMMERCIAL-LIKE AR PILOT PROGRAMS

Examination of these multiple AR pilot programs and extensive interviews with industry experts suggest that significant savings may be achieved in numerous programs if such programs are radically restructured to institutionalize commercial-like incentives for contractors. Nonetheless, the cost savings projections generated by nearly all these programs are not robust and are far from definitive. This is primarily because none are based on actual cost data from the development and manufacture of the specific item in a pre-AR environment compared to the development and manufacture of the same item in a post-AR environment. Indeed, few of these estimates use any "actuals" even for the post-AR number. Instead, all are projections based on expectations, analogous experience, or other factors and are therefore not sufficiently reliable for mathematical cost-estimating models.

Based on the evidence presented in Chapters Four and Five, we conclude that the available projections for cost savings on the majority of existing AR commercial-like pilot programs are not sufficiently robust to support the development of precise technical cost adjustment factors for formal mathematical cost estimating methodologies or CERs for fixed-wing combat aircraft.

This is not meant to imply, however, that commercial-like AR reform measures will not result in cost savings. Indeed, we believe that they likely will. Yet the jury is still out. In addition, many problematic issues are raised by commercial-like AR reform measures, such as the possible need for total system contractor logistics support—issues that require considerable further analysis before a comprehensive judgment can be made.

Given this lack of data and the many uncertainties and complexities that have been cited, how should cost estimators deal with commercial-like AR programs? It is our view that if an acquisition program entails extensive CMI and insertion of commercial (COTS) parts and technology, specific cost reductions need to be assessed as appropriate, most probably at the purchased-material level. For programs such as JDAM and various avionics efforts that claim large savings from AR, vendor-supplied parts, components, boards, and so forth account for up to 80 to 90 percent of recurring costs. Yet there can be wide variations from one system or program to another. Thus, no easy rule of thumb can be applied in this area.

If separate and significant AR initiatives can be identified in specific programs, they should be evaluated individually and the results used to adjust the baseline cost estimate, assuming that the baseline is based on historical, pre-AR costs. One of the most important AR initiatives is the extensive use of CAIV. However, once the final design configuration is determined and frozen following the CAIV process, the AR savings from CAIV should already be clearly reflected in the LCC baseline of the system. Yet if a program entails significant contractor configuration control throughout EMD and production, a careful assessment of ongoing cost-saving opportunities and contractor incentives is warranted. Possible positive and negative O&S implications of contractor configuration and Total System Performance Responsibility (TSPR) need to be examined.

Chapter Six
MULTIYEAR PROCUREMENT

Multiyear procurement entails the purchase by the DoD of more than one year's requirement of a production article with a single contract. Multiyear procurement reduces costs by permitting more efficient use of the contractor's resources. For example, the contractor can plan the purchase of long-lead items and materials for a longer production run and thus gain efficiencies through larger-quantity orders. Long-term hiring and personnel planning can also be made more efficient. Tooling efficiencies are also likely. Finally, avoidance of the nonrecurring costs associated with negotiating and implementing a new contract each year saves money as well.

Multiyear contracts are currently limited to five years and must be explicitly authorized by Congress. The DoD must usually demonstrate potential savings and show an adequate commitment to long-term funding before receiving permission from Congress to proceed with a multiyear contract. Normally, multiyear contracts are used only on relatively mature systems that have low technical risk and are already in production.

Estimates for multiyear contract savings for a selected list of programs and proposals are shown in Table 6.1. Three multiyear contracts have been awarded for the Lockheed (formerly General Dynamics) F-16 fighter program. A fourth was also considered. The estimated cost savings varied between 5.4 and 10 percent.

In 1996, the DoD and Congress approved multiyear funding for a single contract for the remaining 80 Boeing (formerly McDonnell Douglas) C-17 strategic airlifters out of a planned buy of 120. The airframe prime contractor and the engine prime contractor both received seven-year multiyear procurement contracts, as shown in Table 6.1. The DoD also designated the engine program (Pratt & Whitney [P&W] F117) a DAPP so that normal statutory procurement regulations could be waived.

The Javelin is a shoulder-launched anti-tank guided missile (ATGM) combined with a reusable command launch unit (CLU). In September 1996, the Army and the contractor proposed a firm-fixed-price contract based on a three-year multiyear contract. Projected savings were put at more than 14 percent.[1]

The Army-run program for development of a Medium Tactical Vehicle Replacement (MTVR) aims at fulfilling the U.S. Marines' future tactical truck requirements. In February 1999, a five-year multi-year contract for the production of more than 8000 vehicles was awarded to Oshkosh Truck Corporation. In 2000, the MTVR entered the LRIP phase. Estimated savings from the multiyear project stood at 7.4 percent.

For the CH-60 helicopter, the Navy plans to combine its buy with the Army's H-60 multiyear contract to achieve its savings. The CH-60 Fleet Combat Support Helicopter is planned to replace the Navy's aging fleet of H-46 helicopters. Missions of the CH-60 include mine countermeasures operations, combat search and rescue, special operations, and fleet logistics support. The Navy variant will use the existing U.S. Army UH-60 Blackhawk fuselage, which is already in production, and will add various systems from the Navy SH-60 Seahawk. Development of the Navy variant began in FY98, and production is planned to commence in FY99. The Navy requirement is for 237 of these aircraft. Multiyear contracting is expected to save 5.5 percent.

---

[1]GAO was critical of this proposal, arguing that the Javelin was not technologically mature enough to warrant a multiyear procurement contract. See U.S. General Accounting Office, *Army Acquisition: Javelin Is Not Ready for Multiyear Procurement*, GAO/NSIAD-96-199, September 26, 1996.

Table 6.1

Selected Multiyear Procurement Savings Estimates

Program	Production Savings (%)	Source	Estimate Quality
F-16 (FY82–85) (534 aircraft)	10	1	Forecast
F-16 (FY86–89) (720 aircraft)	10	1	Forecast
F-16 (FY90–93) (330 aircraft)	5.5	1	Forecast
F-16 (FY99–02) (48 aircraft, LMTAS only[a])	5.4	1	Forecast
Commercial Derivative Engine (CDE) (P&W F117 for C-17)	8.2 (7 years)	1, 2, 3	Forecast
C-17 (airframe) (7 years)	5.5	1, 3 (contract)	Forecast
Javelin ATGM (FY97–99) (2600, 260 CLUs)	14.3	1, 4	Analysis
Medium Tactical Vehicle Replacement (MTVR)	7.4	5	Analysis
CH-60 (Navy + Army multiyear)	5.5	5	Forecast
DDG-51 Aegis (FY98–01)	9	6	Forecast
F-22 (1996 estimate)	3.9–4.7	7	Analysis
F/A-18E/F (FY00 target)	7.4	8	Analysis

SOURCES:
(1) *Acquisition Reform Cost Savings and Cost Avoidance: A Compilation of Cost Savings and Cost Avoidance Resulting from Implementing Acquisition Reform Initiatives*, AFMC draft report, December 19, 1996.
(2) Office of the Deputy Under Secretary of Defense, Acquisition Reform, Pilot Program Consulting Group, *PPCG 1997 Compendium of Pilot Program Reports*.
(3) Department of Defense, Pilot Program Consulting Group, *Celebrating Success: Forging the Future*, 1997.
(4) U.S. General Accounting Office, *Army Acquisition: Javelin Is Not Ready for Multiyear Procurement*, GAO/NSIAD-96-199, September 26, 1996.
(5) "DoD Navy and Marine Corps Modernization Programs for FY 1999," testimony before the Military Procurement and R&D Subcommittees of the House National Security Committee, March 4, 1999.
(6) Ronald O'Rourke, Congressional Research Service, statement before House Committee on Armed Services, March 9, 1999.
(7) U.S. General Accounting Office, *Tactical Aircraft: Restructuring of the Air Force F-22 Fighter Program*, GAO/NSIAD-97-156, June 4, 1997.
(8) Hearings, Committee on Armed Services, Military Procurement Subcommittee, House of Representatives, March 3, 1999.
(9) Savings percentages include government investments for cost reduction initiatives for C-17 airframe and F/A-18E/F.

[a] LMTAS = Lockheed Martin Tactical Aircraft Systems.

The DDG-51 Arleigh Burke–class guided missile destroyer is one of the Navy's most important multimission surface combatants. Twenty-five have been delivered through FY98, and 20 more are under contract. Congress authorized a multiyear procurement of 12 ships in FY98 through 2001, with claimed savings of $1.4 billion compared to a conventional acquisition strategy.

The F-22 estimates were made in 1996 by the Joint Estimate Team (JET) appointed by the Assistant Secretary of the Air Force for Acquisition and by the DoD CAIG. Different estimates now exist for savings from multiyear procurement funding of the F-22, but they are not presented here.

The Navy hopes to save between 7.4 and 10 percent through a five-year multiyear procurement contract with Boeing for the F/A-18E/F. The 7.4 percent figure is the minimum program savings goal that must be certified to Congress before the multiyear funding program is authorized.

Again, it is important to mention a key caveat regarding the comparisons on which these savings claims (as well as many other savings claims) are made: These claims are based on comparing preprogram estimates of the program costs of a year-to-year contract to a multiyear basis. Once a decision is made to follow one path or the other, the two can no longer be compared on an equivalent basis, as fact-of-life changes occur throughout a production program. Thus, the savings are based on the best estimates available at the time of the decision, not on any "actual" historical data.

Based on the evidence collected here, and keeping in mind the caveats stated above, we conclude that multiyear contracts that are effectively implemented by the prime contractor and government customer can be expected to produce approximately 5 percent or greater savings compared to traditional programs. Multiyear contracts permit long-range planning by contractors. In addition, they permit larger buys of materials and parts and allow for the establishment of strategic relationships between primes and subcontractors. Therefore, multiyear contracting should inherently result in some cost savings. However, strategic sourcing relationships between primes, subcontractors, and suppliers fostered under lean manufacturing will have to be evaluated by cost estimators in conjunction with the multiyear savings to ensure that double counting is avoided.[2]

---

[2] See Cook and Graser, *Military Airframe Acquisition Costs: The Effects of Lean Manufacturing,* for a discussion of strategic supplier relationships.

Chapter Seven
CONCLUSIONS ON AR COST SAVINGS ESTIMATES

This chapter presents tables that summarize the data outlined in this report on AR cost savings estimates. The many assumptions underlying each estimate and the numerous caveats included in the body of this report are not all repeated here. Therefore, the estimates shown here must be viewed with extreme caution and should not be used without a detailed reading of the main text of the report. This is necessary if the reader is to clearly understand the many limitations and caveats that must be applied in the application of these estimates.

With the exception of the first table, all the tables in this chapter show a column labeled "estimate quality." It should be remembered that we made a distinction between three levels of estimate quality. The highest-quality estimate is labeled "actuals." This means that the estimate of AR savings was based on actual R&D and production data from the specific item under consideration compared to earlier actuals for the program prior to the imposition of AR. Virtually none of the estimates available during this research effort was of this type.[1]

The second-highest-quality estimate is labeled "forecast." This type of estimate refers mainly to a narrow set of cases in which actual production costs for the specific article are well known but the program is being restructured in a way that is expected to reduce costs. This applies primarily to estimates of multiyear production contract savings.

---

[1] See further discussion below on "actuals."

The third-highest-quality estimates are labeled "analysis." These estimates are made in situations where no actual costs are available for the specific item. In such cases, the anticipated pre-AR cost of a specific item, which has not yet been fully developed or entered into production, is compared to the expected cost of that item after the imposition of AR. In other words, neither the actual cost of the item under the old system nor the actual cost of the item after the imposition of AR is known. This type of estimate is based on analysis, past experience, data from analogous military or commercial programs, expert opinion, or other such methods.

Almost all the estimates of AR cost savings collected in this report fall into the category of "analogies." That is to say, they are not based on actual data for the specific system or the specific program structure in question. This is another key reason these estimates must be treated with extreme care.

SUMMARY OF THE DoD REGULATORY AND OVERSIGHT COST PREMIUM ESTIMATES

Table 7.1 summarizes early ROM estimates of the DoD cost premium. Some of these estimates include potential cost savings from factors other than the reduction in compliance costs, such as cost benefits from using commercial technologies and parts. Most of these estimates should be characterized as informed guesses.

Table 7.2 summarizes the most important estimates of the DoD regulatory and oversight compliance cost premium based on actual data as developed by Coopers & Lybrand and by other studies using similar approaches and methodologies during the beginning phases of the current AR reform effort.[2] The DoD regulatory and oversight compliance cost premium arises from the cost to contractors of complying with a large number of U.S. government and DoD regulations and reporting requirements that are not applicable in the commercial world. These studies, which employ a variety of different methodologies, used actual data for the baseline and then projected cost savings or cost drivers based largely on expert opinion and data projections rather than on actual post-AR data.

---

[2]The information cutoff date for this document was December 1999.

Table 7.1

Early ROM Estimates of the DoD Regulatory and Oversight Cost Premium[a]

Study	Date	Estimated DoD Cost Premium/ Potential Cost Savings (%)
Honeywell defense acquisition study (20 programs, contractor costs)	1986	13
RAND OSD regulatory cost study (total program costs)	1988	5–10
OTA industrial base study (total DoD acquisition budget)	1989	10–50
CSIS CMI study (cost premium on identical items)	1991	30
Carnegie Commission (total DoD acquisition budget)	1992	40
ADPA cost premium study (product costs)	1992	30–50

[a]The full titles of these studies are as follows: *Defense Acquisition Improvement Study*, Honeywell, May 1986; G. K. Smith et al., *A Preliminary Perspective on Regulatory Activities and Effects in Weapons Acquisition*, Santa Monica: RAND, R-3578-ACQ, March 1988; Office of Technology Assessment, *Holding the Edge: Maintaining the Defense Technology Base*, Vol. II Appendix, Washington, D.C.: USGPO, April 1989; *Integrating Commercial and Military Technologies for National Security: An Agenda for Change*, Washington, D.C.: Center for Strategic and International Studies, April 1991; *A Radical Reform of the Defense Acquisition System*, Carnegie Commission on Science, Technology and Government, December 1, 1992; and *Doing Business with DoD—The Cost Premium*, American Defense Preparedness Association, 1992.

We believe that the most reliable of these studies suggest potential savings from DoD regulatory and oversight relief in the range of 1 to 6 percent. We suggest that this range, with an average of 3.5 percent, is a reasonable rule-of-thumb estimate for potential savings from reducing the DoD regulatory and oversight compliance cost premium.

We also examined nongovernment and GAO studies of overall DoD AR program savings that were undertaken in the early phases of the Clinton administration's AR efforts. Since most of the programs examined for these estimates had been under way for some time as traditional programs, it is not unreasonable to assume that most of the reported actual savings (as opposed to the reported future cost avoidance beyond FY01) were due to reductions in the DoD regulatory and oversight burden. These estimates are summarized in Table 7.3.

Table 7.2
Estimates of the DoD Regulatory and Oversight Compliance Cost Premium

Study or Program and Date	C&L Top 10 Cost Drivers (%)	C&L Top 24 Cost Drivers (%)	Overall Cost Premium or Savings Potential (%)	Estimate Quality
C&L (1994)	8.5	13.4	18	Forecast
NORCOM (1994)			27	Forecast
DoD Regulatory Cost Premium Working Group (1996)		6.3		Forecast
DoD Reinvention Lab (1996)	1.2–6.1			Forecast
SPI (1998)			0.5	Limited actuals
WCMD (1996) (CDRLs only)			3.5 (R&D)	Analysis
FSCATT (1995)			2	Analysis
B-2 Upgrade (CDRLs only)			2.3	Forecast

Table 7.3
Summary of Initial Assessments of Overall DoD AR Savings

Study and Date	FY95–FY01	1996	FY95–FY02	Estimate Quality
RAND (1996)	4.4			Forecast
MIT (1997) (average of 23 MDAPS)		4.3		Forecast
GAO (1997) (average of 33 MDAPs)			−2[a]	Forecast
GAO (1997) (average of 10 MDAPs with cost savings)			4	Forecast

[a]As noted earlier, this estimate does not dispute the existence of cost savings from AR for these programs. Rather, it suggests that on average, cost savings are often offset by cost increases elsewhere or by reinvestment.

Although these numbers are not directly comparable either to each other or to earlier estimates of potential DoD regulatory and oversight reform cost savings, we believe that they add some support to the notion that the DoD regulatory and oversight cost burden is in the range of 1 to 6 percent. This is because most of the savings identified are likely to have resulted from reductions in the DoD regulatory and oversight burden at this time, since more radical programmatic acquisition reforms had not yet been fully implemented.

It is important to note, however, that if these potential savings are to be fully achieved, most of the significant regulations and oversight imposed by the DoD must be removed in their entirety from all programs conducted by a specific contractor or at a specific facility. Otherwise, much of the overhead costs that account for the DoD regulatory and oversight cost premium will remain.

SUMMARY OF SAVINGS FROM AR PILOT AND DEMONSTRATION PROGRAMS (COMMERCIAL-LIKE PROGRAM STRUCTURE)

These programs exhibit a complex mixture of numerous reform measures that are outlined in detail in the body of this report. The purpose of these measures is to structure weapon system acquisition programs in a manner that more closely resembles commercial R&D and production programs that emphasize cost as a primary objective and use commercial standards, technology, parts, and components. Table 7.4 summarizes the cost savings estimates from these programs.

These data suggest that R&D savings in the range of 15 to 35 percent may be possible in certain types of programs that are fully restructured in a commercial-like manner in accordance with concepts of CAIV, as discussed in detail in the body of this report. The likely scale of anticipated production savings is more uncertain. However, the three best-documented cases—JDAM, WCMD, and JASSM—suggest that savings of up to 65 percent could be possible, at least in programs for less complex systems with high production runs.

Table 7.4

Summary of Savings from AR Pilot and Demonstration Programs (Commercial-Like Program Structure)[a]

Program	Program Savings (%)	R&D Savings (%)	Production Savings (%)	Estimate Quality
JDAM		15	60	Forecast
WCMD		35	64	Forecast
JASSM	44[b]	29	31	Analysis
EELV		20–33	25–50	Analysis
SBIRS		15		Analysis
FSCATT	13.5	16–34	7	Analysis
JPATS	18.9[c]	13.6	−26.6	Analysis
Tier III–		20		Analysis
Tier II+		3		Analysis
ASP		30		Analysis
AAAV			10–20	Analysis

[a]Note the important qualifications explained in the text.
[b]Overall program cost savings claimed by the DoD, March 1999.
[c]Despite a large increase in production costs, overall program cost estimates declined significantly because of a large anticipated reduction in O&S costs. In March 1999, the DoD claimed an overall JPATS contract cost savings of 49 percent.

Some serious qualifications must be noted in discussing these outcomes. For example, the reforms implemented on these pilot programs have not been widely used as an integrated package outside these AR demonstration programs. Furthermore, many AR pilot programs are relatively small and are characterized by low technological risk, commercial derivative items, and large production runs. Thus, the scale of potential cost benefits for a large, complex weapon system that employs high-risk, cutting-edge technology remains uncertain. Most significantly, none of these programs have entered into full-rate production, and many have not even entered into full-scale development. Thus, in most cases both the baseline estimates of pre-AR costs and the post-AR estimates on which the savings estimates are based are not founded on actual hard data from the development and production of the specific article.

SUMMARY OF MULTIYEAR PROCUREMENT SAVINGS ESTIMATES

The estimates we collected suggest that multiyear contracts can save 5 percent or more on production contracts. Table 7.5 summarizes these estimates.

CONCLUDING OBSERVATIONS: SOME RULES OF THUMB FOR COST ESTIMATORS

Given the wide diversity of claimed savings arising from AR and the serious doubts raised about the reliability and robustness of many of the estimates in this report, how should cost estimators, program managers, and government personnel involved in financial planning account for savings for new acquisition programs? We would recommend the following rules of thumb for preparing cost estimates for new systems under an AR environment:

- Based on the evidence collected and evaluated in this report, we believe it is reasonable to assume that program savings of 3 to 4 percent will result from reductions in the DoD regulatory and

Table 7.5

Summary of Multiyear Procurement Savings Estimates[a]

Program	Production Savings (%)	Estimate Quality
F-16 (FY82–85)	10	Forecast
F-16 (FY86–89)	10	Forecast
F-16 (FY90–93)	5.5	Forecast
F-16 (FY99–02)	5.4	Forecast
CDE	8.2	Forecast
C-17 (airframe)	5.5	Forecast
Javelin ATGM	14.3	Analysis
MTVR	7.4	Analysis
CH-60 (USN + USA)	5.5	Forecast
DDG-51 (FY98–01)	9	Forecast
F-22 (1996 CAIG/JET)	3.9–4.7	Analysis
F/A-18E/F (target)	7.4	Analysis

[a]Savings percentages include government investments for cost reduction initiatives for C-17 airframe and F/A-18E/F.

oversight burden. In other words, if one is using a pre-AR program (prior to 1994) as an estimating analogy for a similar new program, cost reductions at the program acquisition level of 3 to 4 percent can reasonably be expected to result from reductions in the regulatory and oversight burden, assuming that a comprehensive elimination of the most significant cost-driving regulations and reporting requirements has been implemented (see above).

- However, if the cost analysis is developed using prior program direct or indirect labor hours, most of the AR savings from reductions in regulatory and oversight burden should already be reflected in the negotiated Forward Pricing Rate Agreements (wrap rates), so no further adjustment would be warranted in the rates themselves. This is because most regulatory burden cost savings are in the area of indirect costs and should thus show up in overhead cost savings. Because AR has been in existence since 1995, most of the realizable reductions in regulatory and oversight burdens should already have been calculated between the contractor and the Defense Contract Management Agency. Again, this assumes that facility-wide or contractor-wide regulatory relief has been applied across the board.

- AR reductions between suppliers and the prime may have to be assessed separately. Factors such as regulatory flow-down and the cost effects of strategic supplier relationships need to be taken into account. Although AR focuses mainly on interactions between the government and the primes, there are likely areas between primes, subcontractors, and suppliers that may result in further savings.

- If an acquisition program entails extensive CMI and insertion of commercial (COTS) parts and technology, specific cost reductions need to be assessed as appropriate, probably at the purchased material level. For programs such as JDAM and various avionics efforts that claim large savings from AR, vendor-supplied parts, components, boards, and so forth account for up to 80 to 90 percent of recurring costs. Yet there can be wide variation from one system or program to another. Thus, no easy rule of thumb can be applied in this area.

- If separate and significant AR initiatives can be identified in specific programs, they should be evaluated individually and the results used to adjust the baseline cost estimate, assuming that the baseline is based on historical, pre-AR costs. One of the most important AR initiatives is the extensive use of CAIV. Once the final design configuration is determined and frozen following the CAIV process, the AR savings from CAIV should already be clearly reflected in the LCC baseline of the system. However, if a program entails significant contractor configuration control throughout EMD and production, a careful assessment of ongoing cost-saving opportunities and contractor incentives is warranted. Possible positive and negative O&S implications of contractor configuration and TSPRs need to be examined.

- Based on the evidence collected here, multiyear contracts that are effectively implemented by the prime contractor and government customer can be expected to produce 5 percent or greater savings compared to traditional programs. Strategic sourcing relationships between primes, subcontractors, and suppliers fostered under lean manufacturing will have to be evaluated in conjunction with multiyear savings to ensure that double counting is avoided.[3]

---

[3] See Cook and Graser, *Military Airframe Acquisition Costs: The Effects of Lean Manufacturing,* for a discussion of strategic supplier relationships.

Appendix A
SUBJECTS OF THE THREE RAND STUDIES ON INDUSTRY INITIATIVES DESIGNED TO REDUCE THE COST OF PRODUCING MILITARY AIRCRAFT

MR-1370-AF, *Military Airframe Costs: The Effects of Advanced Materials and Manufacturing Processes*, by Obaid Younossi, Michael Kennedy, and John C. Graser (2001):

- Automated fiber placement
- Computer-aided design/computer-aided manufacturing (CAD/CAM)
- Electron beam (E-beam) curing
- Filament winding
- Infrared thermography
- High-speed machining
- High-performance machining
- Hot isostatic press casting
- Laser forming of titanium
- Laser ply alignment
- Laser shearography
- Laser ultrasonics
- Optical laser ply alignment
- Out-of-autoclave curing

Pultrusion

Resin film infusion

Resin transfer molding

Statistical process control

Stereolithography

Stitched resin film infusion

Superplastic forming/diffusion bonding

Unitization of aircraft structure

Ultrasonic inspection

Vacuum-assisted resin transfer molding

MR-1329-AF, *An Overview of Acquisition Reform Cost Savings Estimates,* by Mark A. Lorell and John C. Graser (2001):

Civil-military integration

Commercial-like programs

Commercial insertion

Commercial off the shelf (COTS)

Contractor configuration control

Cost as an independent variable (CAIV)

Defense Acquisition Pilot Programs (DAPPs)

Federal Acquisition Reform Act (FARA)

Federal Acquisition Streamlining Act (FASA)

Integrated Product Teams (IPTs)

Military specification reform

"Must-cost" targets

Multiyear procurement

"Other Transactions" Authority (OTA)

Procurement price commitment curve

Regulatory and oversight burden reductions

Single-Process Initiative (SPI)

MR-1325-AF, *Military Airframe Acquisition Costs: The Effects of Lean Manufacturing*, by Cynthia R. Cook and John C. Graser (2001):

Cellular manufacturing

Computer-aided design/computer-aided manufacturing (CAD/CAM)

Continuous flow production

Design for manufacturing and assembly (DFMA)

Electronic data interchange (EDI)

Electronic work instructions (EWI)

Enterprise resource planning (ERP)

First-time quality

Flexible tooling

Integrated Product Teams (IPTs)

Just-in-time (JIT) delivery

Kaizen events

Kitting of parts or tools

Lean Aerospace Initiative (LAI)

Lean enablers

Lean human resource management (HRM)

Lean pilot projects

Operator self-inspection

Production cost reduction plans (PCRPs)

Purchasing and supplier management (PSM)

Pull production

Single-piece flow production

Six-sigma quality
Six "S's" of housekeeping
Statistical process control (SPC)
Strategic sourcing agreements
Takt time
Target costing
Three-dimensional (3D) design systems
Total Productive Maintenance (TPM)
Unitization/part count reduction
Visual manufacturing controls (*Kanban*)
Value (cost) stream analysis

Appendix B
ACQUISITION REFORM COST QUESTIONS

For analytical purposes, AR cost savings initiatives can be divided into three basic categories, as shown below: (1) regulatory/oversight relief; (2) CMI; and (3) requirements reform. (We make a distinction between AR and best business practices; the latter is covered elsewhere.) Separate from these AR cost savings categories are a variety of implementation measures and incentives that promote a new and more cooperate government-contractor relationship. These measures may also have cost implications. On the following two pages we lay out the DoD's basic AR principles, our AR cost savings categories, and some AR implementation measures. Our basic questions are:

1. In your experience with military aircraft or other weapon system programs, what cost savings—if any—are attributable to each of the cost savings categories listed below (regulatory/oversight relief, CMI, requirements reform)? How would you rank their importance and their relative contribution to cost savings? Is it possible to provide hard cost data or other evidence to demonstrate these cost savings?

2. If this level of detail is not possible, what cost savings—if any— are attributable to AR in general? Is it possible to provide hard cost data or other evidence to demonstrate these cost savings?

3. Some of these cost savings initiatives and approaches may require increased up-front program costs or other types of implementation costs. For example, by deemphasizing mil specs, contractors must spend money to qualify COTS parts and components for military applications. Also, CAIV may require far

more operational analysis and trade studies by the contractor. The rationale has been that these approaches produce much greater savings downstream, which compensates for the increased up-front costs. Has this been your experience? Do you have cost data to back up your answers?

4. One of the most detailed studies of the benefits of AR—the well-known Coopers & Lybrand study—claims that the regulatory cost premium (Category 1 below) on military contracts over commercial contracts averages 18 percent (16 percent for the aerospace industry). Has this been your experience? Do you have cost data to back this up? Category 1 also lists the top ten regulatory cost drivers for the aerospace industry as determined by Coopers & Lybrand. Does this list conform with your experience?

BASIC PRINCIPLES

Develop a more "commercial-like" defense procurement process:

- Reduce regulatory burden, decrease government oversight/control.
- Transfer more program, cost, design, and technology control authority and responsibility to the contractor.
- Exploit commercially developed parts, components, technologies, and processes.
- Make cost/price a key system requirement.

AR COST-SAVING CATEGORIES

1. Regulatory/oversight relief[1]
    - Cost/schedule reporting and accounting reform
        - TINA waivers/reform (#2)

---

[1] With the Coopers & Lybrand/TASC 1994 "Key Cost Drivers" ranking for the aerospace industry sector. The top ten represent 60.7 percent of the total DoD premium cost for aerospace of 16 percent.

- C/SCSC reform (#3)
- DCAA audits/DCMC interface reform (#6)
- CAS reform
- Specifications and standards reform
  - Quality program (MIL-Q-9858A) reform: ISO 9000 (#1)
  - Configuration management system (MIL-STD-973) reform (contractor configuration control) (#4)
  - Engineering drawings (MIL-STD-100) reform (commercial standards)
  - Single-Process Initiative
- Contracting, purchasing, materials/property management reform
  - Program reviews (#5)
  - Government property administration reform (#7)
  - Government facilities mods (FAR Subpart 36) (#8)
  - Contractor Purchasing Requirements, including Contractor Purchasing System Review (CPSR) (#9)
  - Contract clauses, superfluous requirements/SOW reform (SOO) (#10)
  - MMAS reform
2. Commercial-Military Integration (CMI)
   - Mil spec/standards elimination/reform
   - Commercial insertion (use of less expensive COTS or ruggedized commercial parts, components, designs, architectures, technologies)
     - Commercial Operations and Support Savings Initiative (COSSI)
   - Dual-use R&D and production: economies of scale, military production on commercial lines

- TRW/CNI (communications, navigation, identification–friend or foe)—Mantech

3. Requirements reform
   - Prioritized mission performance requirements instead of detailed technical specs; contractor develops technical solution, "owns" design and technical approach
     - Focus on critical military performance requirements: reject gold plating
   - CAIV
     - Treat cost or price as a variable of equal or greater importance compared to performance and schedule when conducting trade-off analyses throughout the design and R&D phases of the program
     - CAID (Clear Accountability in Design)
     - No mil specs required
     - Encourage use of commercial parts, components, processes, technologies

ACQUISITION REFORM IMPLEMENTATION APPROACHES

Below are some popular AR implementation initiatives and measures affecting the government-contractor relationship and involving incentives, competition, and partnering. In your view, do these measures promote AR and cost savings? What are the cost implications of these measures? Can you provide data to back up your answers?

- Government-industry IPPTs: maximum sharing of info
- R&D risk reduction phases with multiple contractors
- Contractor configuration control
  - Grant contractor partial or full control over design, configuration, technical solutions (configuration control/ change authority for second/third levels)
- Maintaining competition as long as possible; combined with rolling down-select

- Frequent updates to contractors on their relative standing in the competition and why
- Emphasis on past performance criteria in down-select
- AUPPC curves from contractors with carrot/stick incentives
  - Contractual carrot to contractor: No cost auditing, configuration control, guaranteed single source if meets AUPPC, no data package. Contractor keeps additional profit if costs reduced
  - Contractual stick to contractor: mil spec data package to government, mil spec cost auditing rules, must qualify second-source contractor if fails to meet AUPPC
- Multiyear funding/program stability
- Warranties, nonperformance penalties
- OEM contractor logistics support

BIBLIOGRAPHY

Acquisition Reform Benchmarking Group, *1997 Final Report*, Under Secretary of Defense, Acquisition and Technology, June 30, 1997.

Acquisition Reform Cost Savings and Cost Avoidance: A Compilation of Cost Savings and Cost Avoidance Resulting from Implementing Acquisition Reform Initiatives, AFMC draft report, Wright Patterson AFB, Dayton, OH, December 19, 1996.

Activity-Based Cost Analysis of Cost of DoD Requirements and Cost of Capacity: Executive Summary, NORCOM, May 1994.

Anderson, Michael H., *A Study of the Federal Government's Experiences with Commercial Procurement Practices in Major Defense Acquisitions*, Cambridge, MA: Massachusetts Institute of Technology, June 1997.

Apple press release, "Apple Accelerates Marketshare Strategy, Rolls Out Competitively Priced Macintosh Models," October 21, 1993.

"Approval of Extended JASSM EMD Program Seen Imminent," *Aerospace Daily*, November 11, 1998.

Army Acquisition Reform Newsletter, Issue 6, January 1996.

Assistant Secretary of the Air Force, Acquisition, *JDAM—The Value of Acquisition Streamlining*, no date.

"Beech Aircraft Gets Trainer Contract," *Air Force News*, February 1996.

"Boeing Presses 500-Pound JDAM Kit for U.S., International Buyers," *Aerospace Daily*, September 22, 1998.

"Boeing to Deliver Additional JDAMs Two Months Early," *Aerospace Daily*, May 3, 1999.

Capaccio, Tony, "Lockheed Besting Boeing in Funds," *Denver Post*, March 3, 2000.

Chapman, Suzann, "JASSM Competitors Chosen," *Air Force Magazine*, August 1996.

"Competing JASSM Contractors Chosen," *Air Force News*, June 1996.

Cook, Cynthia R., and John C. Graser, *Military Airframe Acquisition Costs: The Effects of Lean Manufacturing*, Santa Monica: RAND, MR-1325-AF, 2001.

Defense Acquisition Improvement Study, Honeywell, May 1986.

Defense Contract Management Command, *Single Process Initiative Implementation Summary*, October 9, 1998.

Department of Defense, *Defense Acquisition Handbook*, June 30, 1998.

Department of Defense, Office of the Director, Operational Test and Evaluation, "Joint Direct Attack Munition (JDAM)," *FY97 Annual Report*, February 1998.

Department of Defense, Office of the Director, Operational Test and Evaluation, *DOT&E FY98 Annual Report*.

Department of Defense, Pilot Program Consulting Group, *Celebrating Success: Forging the Future*, 1997.

"DoD Navy and Marine Corps Modernization Programs for FY 1999," testimony before the Military Procurement and R&D Subcommittee of the House National Security Committee, March 4, 1999.

The DoD Regulatory Cost Premium: A Quantitative Assessment, Coopers & Lybrand/TASC, December 1994.

Doing Business with DoD—The Cost Premium, Washington, D.C.: American Defense Preparedness Association, 1992.

Drezner, Jeffrey A., and Geoffrey Sommer, *Innovative Management in the DARPA HAE UAV Program: Phase II Experience*, MR-1054-DARPA, December 1998.

Eisman, Mel, Jon Grossman, Joel Kvitky, Mark Lorell, Phillip Feldman, Gail Halverson, and Andrea Mejia, *The Cost of Future Military Aircraft Avionics Systems: Cost Estimating Relationships and Cost Reduction Initiatives*, Santa Monica: RAND, limited document, not for public distribution, 2001.

"First WCMD B-52 Test Prepared for Continued Development Testing," *Aerospace Daily*, April 29, 1998.

Fulghum, David A., "Boeing Unveils Stealth Standoff Missile Design," *Aviation Week & Space Technology*, March 9, 1998.

"GAO Finds No Reason to Terminate JASSM," *Aerospace Daily*, September 29, 1998.

Greaves, Samuel A., *The Evolved Expendable Launch Vehicle (EELV) Acquisition and Combat Capability*, Air Command and Staff College, Maxwell AFB, AL, March 1997.

Integrating Commercial and Military Technologies for National Security: An Agenda for Change, Washington, D.C.: Center for Strategic and International Studies, April 1991.

"JASSM Beats Cost Target by 40%," *Aerospace Daily*, April 30, 1998.

"JASSM Cruise Missile Crashes in First Flight Test," *Aerospace Daily*, April 21, 1999.

"JASSM Schedule Slip Costs $53 Million," *Aerospace Daily*, August 31, 1999.

"JASSM Takes First Step in Flight Testing," *Jane's Defence News*, August 25, 1999.

"JPATS Contract Awarded," *Air Force News*, June 1995.

King, Chris, "Joint Air-to-Surface Standoff Missile (JASSM): Acquisition Reform in Action," unpublished briefing, undated.

Kuhn, Tom, "EELV: A New Rocket for the Millennium," *Airman*, March 1999.

Leonard, R. S., J. A. Drezner, and G. Sommer, *The Arsenal Ship Acquisition Process Experience: Contrasting and Common Impressions from the Contractor Teams and Joint Program Office*, Santa Monica: RAND, MR-1030-DARPA, October 1998.

Lorell, Mark, Julia Lowell, Michael Kennedy, and Hugh Levaux, *Cheaper, Faster, Better? Commercial Approaches to Weapons Acquisition*, Santa Monica: RAND, MR-1147-AF, 2000.

Moody, Jay, *Achieving Affordable Operational Requirements on the SpaceBased Infrared System (SBIRS) Program: A Model for Warfighter And Acquisition Success?* Air Command and Staff College, Maxwell AFB, AL, March 1997.

"Navy Wants Upgraded JDAM for No More Than $50,000," *Aerospace Daily*, August 26, 1998.

Office of the Assistant Secretary of the Air Force, Acquisition, *Acquisition Reform Success Story: Joint Direct Attack Munition (JDAM)*, December 1997.

Office of the Assistant Secretary of the Air Force, Acquisition, *Acquisition Reform Success Story: Wind Corrected Munitions Dispenser (WCMD)*, June 12, 1997.

Office of the Deputy Under Secretary of Defense, Acquisition Reform, Pilot Program Consulting Group, *PPCG 1997 Compendium of Pilot Program Reports.*

Office of the Deputy Under Secretary of Defense, Acquisition Reform, *Cost as an Independent Variable: Stand-Down Acquisition Reform Acceleration Day*, May 1996.

Office of the Deputy Under Secretary of Defense, Acquisition Reform, *Single Process Initiative, Acquisition Reform Acceleration Day Stand-Down*, 1996.

Office of the Director, Program Analysis and Evaluation, Cost Analysis Improvement Group, Memorandum for the Chairman, Conventional Systems Committee, *CAIG Report on the Life Cycle Costs of the Joint Direct Attack Munition (JDAM) Program*, August 25, 1993.

Office of Technology Assessment, *Holding the Edge: Maintaining the Defense Technology Base*, Vol. II Appendix, Washington, D.C.: USGPO, April 1989.

Office of the Under Secretary of Defense, Acquisition and Technology, Acquisition Reform Benchmarking Group, *1997 Final Report*, June 30, 1997.

Office of the Under Secretary of Defense, Acquisition and Technology, Acquisition Reform Senior Steering Group, DoD Regulatory Cost Premium Working Group, *Compendium of Office of Primary Responsibility (OPR) Reports*, June 30, 1995.

Office of the Under Secretary of Defense, Acquisition and Technology, Acquisition Reform Senior Steering Group, DoD Regulatory Cost Premium Working Group, *Updated Compendium of Office of Primary Responsibility (OPR) Reports*, June 1996.

Office of the Under Secretary of Defense, Acquisition and Technology, *Overcoming Barriers to the Use of Commercial Integrated Circuit Technology in Defense Systems*, October 1996.

O'Rourke, Ronald, Congressional Research Service, statement before the House Committee on Armed Services, March 9, 1999.

Perry, William J., Secretary of Defense, *Acquisition Reform—Mandate for Change*, February 1994.

Perry, William J., Secretary of Defense, *Common Systems/ISO-9000/Expedited Block Changes*, December 6, 1995.

Perry, William J., Secretary of Defense, *Specifications and Standards—A New Way of Doing Business*, June 29, 1994.

"Problems Force Delay in JDAM Full-Rate Production," *Aerospace Daily*, December 15, 1997.

A Radical Reform of the Defense Acquisition System, Carnegie Commission on Science, Technology, and Government, December 1, 1992.

Reig, Raymond W., "Baselining Acquisition Reform," *Acquisition Review Quarterly*, Winter 2000.

Rosensteel, Thomas E., *An Implementation of Cost as an Independent Variable (CAIV): The Space Based Infrared Systems (SBIRS) High Component Program*, SBIRS Program Office, no date.

Rush, Benjamin C., "Cost as an Independent Variable: Concepts and Risks," *Acquisition Renew Quarterly*, Spring 1997.

"SBIR Low Teams Put Finishing Touches on Revised Proposal," *Aerospace Daily*, April 6, 1999.

Schank, John, Kathi Webb, Eugene Bryton, and Jerry Sollinger, "Analysis of Service-Reported Acquisition Reform Reductions: An Annotated Briefing," Santa Monica: RAND, unpublished research, September 1996.

Schwartz, Karl, "Global Hawk Convinces U.S. Air Force," *Flug Revue Online*, November 2000.

Smith, Giles K., Jeffrey A. Drezner, William C. Martel, J. J. Milanese, W. E. Mooz, and E. C. River, *A Preliminary Perspective on Regulatory Activities and Effects in Weapons Acquisition*, Santa Monica: RAND, R-3578-ACQ, March 1988.

Sommer, G., G. K. Smith, J. L. Birkler, and J. R. Chiesa, *The Global Hawk Unmanned Aerial Vehicle Acquisition Process: A Summary of Phase 1 Experience*, Santa Monica: RAND, MR-809-DARPA, 1997.

Statement by the Honorable Jacques S. Gansler, Under Secretary of Defense, Acquisition and Technology, to the Subcommittee on Acquisition and Technology, Committee on Armed Services, U.S. Senate, March 18, 1998.

Statement of Stan Z. Soloway, Deputy Under Secretary of Defense, Acquisition Reform, House Armed Services Committee, March 2, 1999.

Statement of the Under Secretary of Defense for Acquisition and Technology, Honorable Paul G. Kaminski, Before the Acquisition and Technology Subcommittee of the Senate Committee on Armed Services on Defense Acquisition Reform, Committee on Armed Services, U.S. Senate, March 19, 1997.

Sweetman, Bill, "Coming to a Theatre Near You," *Interavia Business and Technology,* July 1999.

"USAF Approves WCMD for Low Rate Production," *Aerospace Daily,* August 4, 1998.

"USAF Has Fix for One WCMD Problem," *Aerospace Daily,* February 20, 1998.

"USAF Postpones JASSM Decision Until 2001," *Aerospace Daily,* August 30, 1999.

"USAF to Begin Planning JASSM Upgrades," *Aerospace Daily,* April 29, 1998.

"USAF Will Buy Only Mk.84 JDAMs This Year," *Aerospace Daily,* December 17, 1997.

U.S. Air Force, *Single Acquisition Management Plan for the Joint Direct Attack Munition (JDAM),* Eglin Air Force Base, August 23, 1995.

U.S. Air Force, *Space-Based Infrared System Fact Sheet,* no date.

U.S. General Accounting Office, *Acquisition Reform: DoD Faces Challenges in Reducing Oversight Costs,* GAO/NSIAD-97-48, January 1997.

U.S. General Accounting Office, *Acquisition Reform: Effect on Weapon System Funding,* GAO/NSIAD-98-31, October 1997.

U.S. General Accounting Office, *Acquisition Reform: Efforts to Reduce the Cost to Manage and Oversee DoD Contracts,* GAO/NSIAD-96-106, April 1996.

U.S. General Accounting Office, *Army Acquisition: Javelin Is Not Ready for Multiyear Procurement,* GAO/NSIAD-96-199, September 26, 1996.

U.S. General Accounting Office, *Evolved Expendable Launch Vehicle: DoD Guidance Needed to Protect Government's Interest*, GAO/NSIAD-98-151, June 1998.

U.S. General Accounting Office, *Precision-Guided Munitions: Acquisition Plans for the Joint Air-to-Surface Standoff Missile*, GAO/NSIAD-96-144, June 1996.

U.S. General Accounting Office, *Tactical Aircraft: Restructuring of the Air Force F-22 Fighter Program*, GAO/NSIAD-97-156, June 4, 1997.

Wall, Robert, "Lingering Concerns Stalk F/A-18E/F," *Aviation Week & Space Technology*, February 28, 2000.

"Wind Corrected Munitions Dispenser Price Holds Despite Fixes," *Aerospace Daily*, March 23, 1998.

Younossi, Obaid, Michael Kennedy, and John C. Graser, *Military Airframe Costs: The Effects of Advanced Materials and Manufacturing Processes*, Santa Monica: RAND, MR-1370-AF, 2001.